COLLAPS

IBLE

THE GENIUS OF SPACE-SAVING DESIGN

by Per Mollerup

COL

LAPSIBLE

CHRONICLE BOOKS
SAN FRANCISCO

COLLAPSIBLE

First published in the United States in 2001 by Chronicle Books LLC

Library of Congress Cataloging-in-Publication Data available.

ISBN: 0-8118-3236-8

Printed in Hong Kong

Design by Mollerup Designlab, Copenhagen

Cover design by Benjamin Shaykin

Distributed in Canada by
Raincoast Books
9050 Shaughnessy Street
Vancouver, BC V6P 6E5

10 9 8 7 6 5 4 3 2 1

Published by
Chronicle Books LLC
85 Second Street
San Francisco, CA 94105

www.chroniclebooks.com

SUPPLIERS

Manufacturers mentioned in *Collapsible* are—to the extent that it has been possible to identify them—current manufacturers, and did not necessarily introduce the product to the market. The inclusion of a product does not imply that it is in current production. When it is not, either the first (preferably) or the most recent manufacturer is given.

ACKNOWLEDGMENTS

Designfonden and Konsul George Jorck og Hustru Emma Jorck's Fond, both in Copenhagen, have generously supported research and photography for this book.

New photographs were taken by Ole Woldbye & Pernille Klemp, Jørgen Schytte, Søren Kvistgaard, and the author. Many designers from Mollerup Designlab were inevitably involved in the birth of *Collapsible*. Jacob Fløche was an unfailing source of inspiration and research. Davood Banissi, Kamilla Bay, Mette Gadegaard, Michael Jerndorff, Mette Kryger, Tenna Nørgaard and Søren Olsen contributed with research, drawings, typographical advice and action. Many other friends helped the project in many other ways. I am grateful to all.

—Per Mollerup

Contents

Man, himself a collapsible being,
physically and psychologically,
needs and wants collapsible tools.

Introduction: A strategy for survival

This is the first general introduction to collapsibles – objects that, in one way or another, fold out for action and fold up for storage. 'Collapsible' used as a noun is a neologism. This book is the first to use the word to refer to any object that is in some way collapsible, rather than to describe it in terms of that quality. Sometimes you need the word first before you can think about a thing in depth.

Collapsibility is an elementary design principle applied to a great many everyday objects, from telescopes to umbrellas, newspapers to Venetian blinds, furniture to perambulators. Collapsibles work by adjustment, which is a basic strategy for survival. The principle is simple: no adjustment, no future. Adapt and survive.

The prime purpose of this book is to discuss a subject never thoroughly explored before. But beyond that, the aim is to inspire designers to create space-saving objects of great elegance and functionality. Mies van der Rohe's 'less is more' and Dieter Rams's 'less is better' have never been so true as they are today.

The selection of objects illustrated in *Collapsible* is heavily biased towards ingenuity. Brilliant solutions have shamelessly been given priority over the satisfaction of commonplace needs, with no attempt at completeness. *Collapsible* is not a manual but an album for inspiration.

Cartoon collapsibles. Artists are often first with the good ideas: some Disney inventions have real-world relevance. *The Terror of the River,* Walt Disney, 1946. © Disney Enterprises, Inc.

Collapsible is divided into three main sections:
Part 1, Fold it, discusses the concept of collapsibility.
Part 2, Mechanics, presents twelve collapsibility principles.
Part 3, Furniture, documents the variety of collapsibility principles in one broad field of application.

1. Fold it

A category is born

◄
A fire engine without collapsible ladder and hose would be of little use in an urban environment.

It is not the strongest of the species that survives, nor the most intelligent; it is the one that is most adaptable to change.
Charles Darwin

Collapsibles are smart man-made objects with the capacity to adjust in size to meet a practical need. They are functional doubles with two opposite states, one folded and passive, one (or more) unfolded and active. They grow and shrink, expand and contract, according to functional need.

Size adjustment to meet functional requirements is a time-honoured principle in nature too, as all real men and their happy mates will confirm. Animals downsize to hide, relax, rest and protect themselves and upsize to brag, threaten, fly, fight and court. Survival of the fittest means survival of the best adjusted. Size adjustment is an essential evolutionary strategy.

Collapsibles are ubiquitous. In fact you are studying one now: this book occupies less space when closed and stacked on your bookshelf than when opened on your table. Your spectacles are a collapsible, as is your handkerchief. Your chair may fold away and your lamp retract.

Collapsibles are by definition also expandables. But to qualify as a true collapsible, an object must also be *repeatedly* collapsible and expandable. One-night stands do not qualify; encores are mandatory.

Man-made collapsibles don't just happen. Two conditions must be met before one is conceived and created. First, somebody must see an advantage in reducing the size of a tool when it is not in use. And second, it must be mechanically possible to reduce the size of that tool.

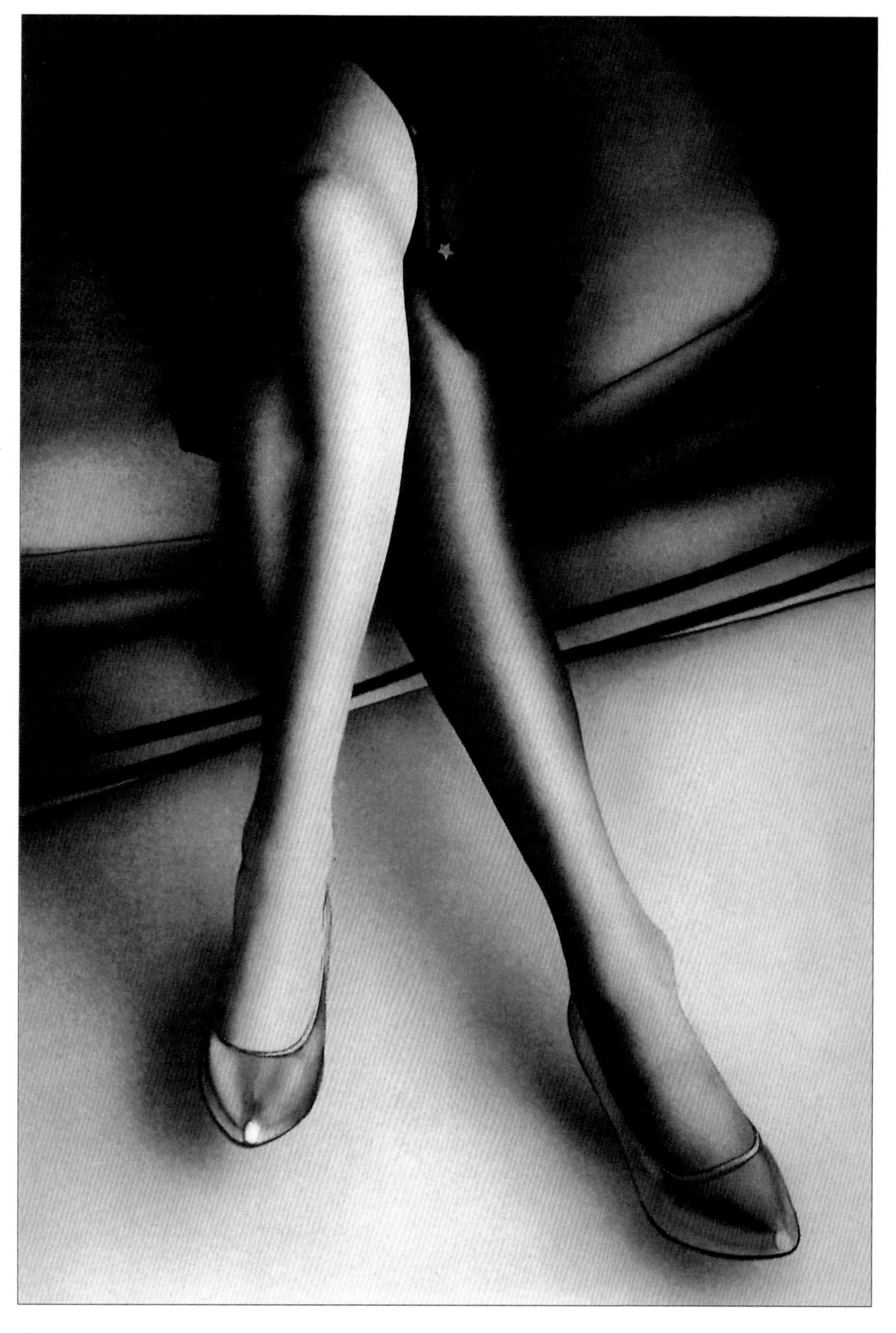

Some collapsibles expand by several hundred per cent while improving drastically in appearance. The US company DuPont introduced nylons – named after New York + London – to the market in 1940. *Alone in the House*, 1980, a watercolour by Hans Henrik Lerfeldt.

A collapsible wicker chair for the motoring picnic. Folded down, it can be carried like a handbag, while the space under the seat holds cushions for seat and back. This generic type was produced in England in the early 1930s.

What we call Venetian blinds in fact originated in Persia. Travellers from 12th-century Venice enjoyed their slatted shade and brought the principle back to their prospering city state. When the British borrowed the idea in the first half of the 20th century they knew no better than to name them Venetian blinds. This S-wave example, manufactured by Art Andersen & Copenhagen, was designed by Lars Mathiesen in 1991.

Tools that perform one function when collapsed and quite another when expanded are an especially clever type of collapsible. The frame of this angler's knapsack opens out into a seat. Manufactured by ABU Garcia, Sweden.

Space saving

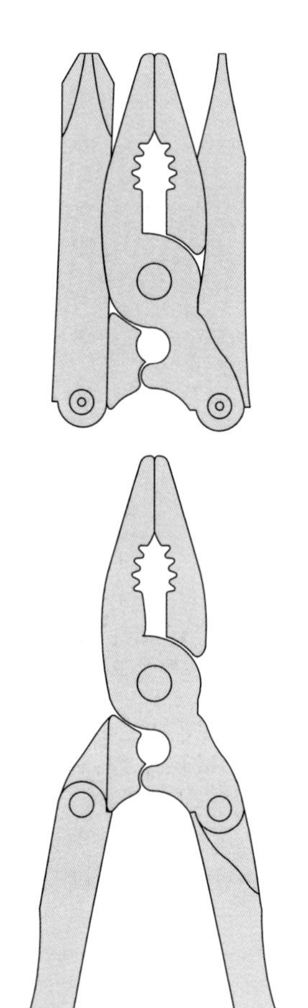

Miniature *and* collapsible (the drawing is actual size), this pair of pliers has a screwdriver on each folding handle. Manufactured by SerberTech, US.

To qualify for collapsibility, an object must have an impractical shape and size – like a newspaper – and maybe a shape which occupies a lot of empty space – like an erected tent. Nobody would think of collapsing a postage stamp or a car engine. A stamp couldn't be more conveniently compact; a car engine is bulky but its form does not leave unused space.

To give collapsibility to an object, the space it occupies – or more precisely, its volume – must be redistributed in one way or another. With the possible exception of objects that are compressed (see p. 32), volume does not actually diminish through collapsing, though it may seem to. If a newspaper is folded a couple of times it certainly *appears* to get smaller, but of course it does not. It merely gains a more practical and portable format. Its volume is *redistributed* so that it occupies less practical space.

Another example: if two stacking chairs are placed one on top of the other, they occupy less practical space than two unstacked chairs. But their volume is unchanged. If – as Archimedes might have done – we lowered the two chairs into a full bath tub, stacked or not they would displace the same amount of water.

Practical space is space we want to free up for some other purpose. In this book, 'space saving' and 'size reduction' mean the redistribution of an object's volume in order to reduce the practical space that it occupies. Collapsibility is above all a measure of convenience.

Danish hunter and author Martin Johnson and memsahib Osa relax on safari in the 1920s, sitting on collapsible furniture in their collapsible home from home. Now, as then, a tent is a house for man on the move. Carried on truck, back or bike, it takes up little practical space. Expanded and erected in the outdoors, it protects its inhabitants from the more troublesome aspects of the natural environment they set out to enjoy.

◄

An umbrella is a man-made adaptation to changes in the weather. *Sudden Shower at Öhasi*, woodcut by Utagawa Hiroshige (1797–1858).

The *chapeau claque* is a genuine collapsible which folds flat for storage and folds out for use. *Chapeau* is French for 'hat'; the word *claque* is onomatopoeic. It echoes the sound made by the hat as it opens with the help of its built-in spring.

Change, mother of invention

The world is in a state of flux. Change is happening all the time, all around us. We try to hang on by continually adapting ourselves and our belongings to our shifting circumstances. When the weather forecast predicts showers, we take a folded umbrella. If it starts to rain, we unfold that umbrella. When the rain stops, we shake it off and fold it up again.

Collapsibles are man-made accommodations to change. Collapsible umbrellas meet meteorological changes; collapsible chairs meet social changes; collapsible tape measures meet the change between the occasional need to take a measurement and the need for convenient storage.

The purpose of redistributing volume is economy: economy of practical space and economy of transportation. How could fire engines move through the streets if ladders could not be folded and hoses rolled? How large would a fire station have to be? Were it not for collapsibility, life and property would be lost.

Nature's own method

◄ Size matters, in animal as much as in human affairs.

Courting male peacocks expand to impress. Their tails may measure a metre or more.

The Venus flytrap opens out to lure its prey, then snaps shut to kill and digest it.

Practical experience shows us that organizations and businesses that fail to adjust to an ever-changing environment tend to disappear. The ability to adapt is necessary to continued survival. So it is in nature.

Many species rely on dynamic adjustment in form and size to reproduce, feed or protect themselves. Animals that hide tend to make themselves small and invisible; aggressors do the opposite. And there are many unique adaptations: the threatened hedgehog makes itself not invisible but unpalatable.

Birds reduce or expand themselves to different sizes when resting, swimming, walking, flying and mating. Courting peacocks strut their stuff with their tails unfurled. Mammals too expand for courtship and mating. Snails shrink to withdraw into their mobile shelters but expand and emerge again when it is safe to venture forth.

Some flowers open and close repeatedly. Open, they attract pollinating insects; closed, they minimize evaporation, maintain a more stable temperature, or temporarily retain an insect to ensure the transfer of pollen. A few carnivorous plant species open out to trap insects for food, then snap shut their 'jaws' around an unsuspecting prey.

The function of collapsibility

Man is a tool-using animal…
Without tools he is nothing,
with tools he is all.
Thomas Carlyle, 1833

All man-made artifacts with a practical function are tools. They are extensions of man's natural capacities. With tools he can accomplish so much more than without.

Collapsibility *per se* is never the purpose function – the *raison d'être* – of a tool. It is always a support function. Some tools have two or more purpose functions: the Ottakringer ladder-chair (opposite) is a good example.

That a tool is collapsible may be very useful or even necessary. If an umbrella did not fold away, it might seem more convenient to risk getting wet. But still the prime reason that we need a tool is its purpose function: a knife to cut with, a chair to sit on, an umbrella to keep the rain off. The reason that we select a specific tool over others, however, may very well be its support functions: a folding knife that won't do damage in the pocket; a deckchair that folds away for storage. Some tools possess an array of support functions, fulfilling multiple secondary requirements. A car has scores of tools that perform support functions – windscreen wipers, headlamps, indicators, remote locking – but it's no use if it won't drive.

Very often any number of generic products will satisfy a buyer's primary requirement, but one in particular will most successfully fulfil a secondary requirement. Collapsibility may never be the most important function of a tool, but it is often the decisive factor when the buyer makes a choice.

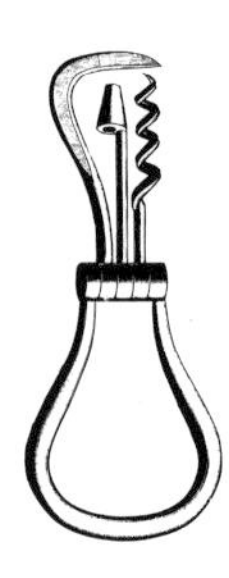

Three in one: folding hoofpick, punch and corkscrew, from an 1890s C.M. Moseman & Brother catalogue, US.

The Ottakringer ladder-chair was inspired by a 19th-century model found in a monastery library in Tyrol, Austria. Ottakringer is the name of the Vienna suburb where it is manufactured today by Section N. The Ottakringer has two purpose functions, as ladder and as chair. Whether it is a collapsible is debatable, for it occupies about the same amount of practical space in both roles.

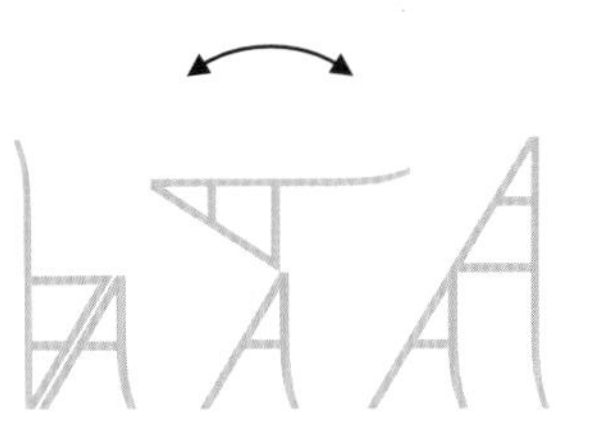

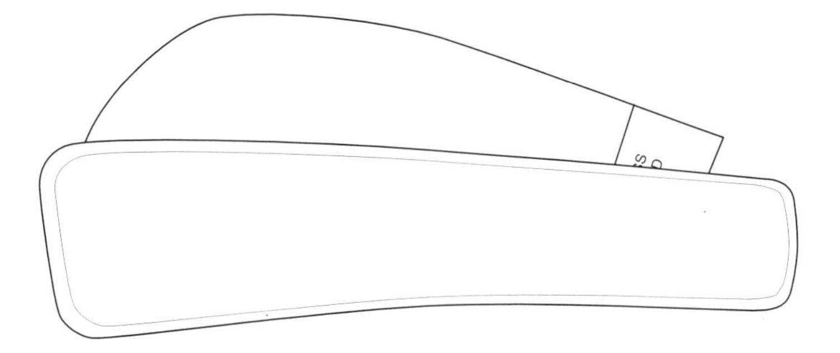

The purpose function of a gardener's pruning knife is to prune – to trim foliage to improve shape or encourage growth. Folding is a support function. Manufactured by Wilkinson Sword, UK.

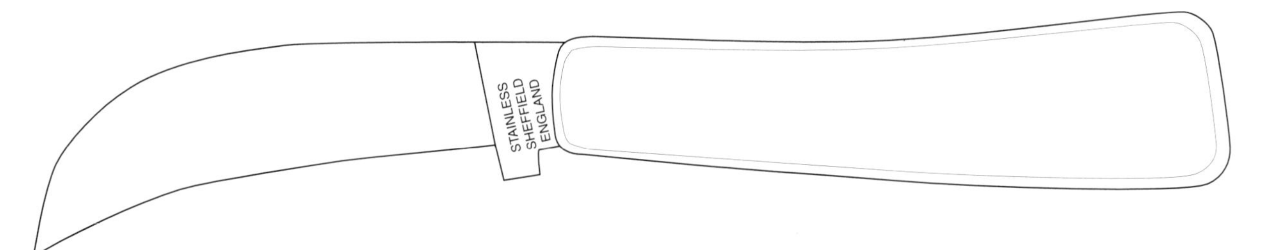

The nons and the quasis

This section is about what collapsibles are not. It deals with classes of objects which, although adjustable in size, do not qualify – or qualify only partially – as collapsibles according to the definition offered here.

– The nons

Objects designed to fold or unfold only once in their lifetime do not qualify as true collapsibles. That excludes one-timers from aerosol-canned shaving foam and popcorn to hand grenades.

This rule also excludes self-assembly or knock-down furniture, for, in practice, once 'unfolded' it is hardly ever 'folded' again. The idea of self-assembly furniture is that it is delivered in pieces to the end customer, who then assembles it himself. The rationale behind this delegation of work is a reduction of costs of manufacture and – primarily – of storage and transportation. To be precise, self-assembly furniture is not collapsed when delivered; it is just not yet fully manufactured.

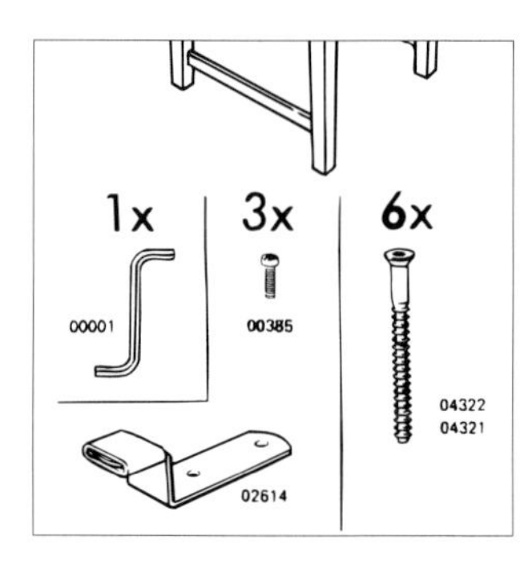

The allen key – number 00001 on the user's instructions – has been IKEA's key to commercial success for half a century. But self-assembly furniture is not collapsible in our sense: it is not designed to 'fold' and 'unfold' again and again.

The disposable paper Dixie Cup was designed in 1908 to replace public metal drinking cups in US trains and public buildings. It was initially called the 'Health Kup', after scares about contamination from common drinking vessels, but in 1919 was renamed the 'Dixie Cup' to give it stronger brand identity in the face of growing competition. Although paper Dixie Cups (and their plastic successors) 'fold' collectively – that is, they stack – they are not intended for reuse, so they don't count as collapsibles.

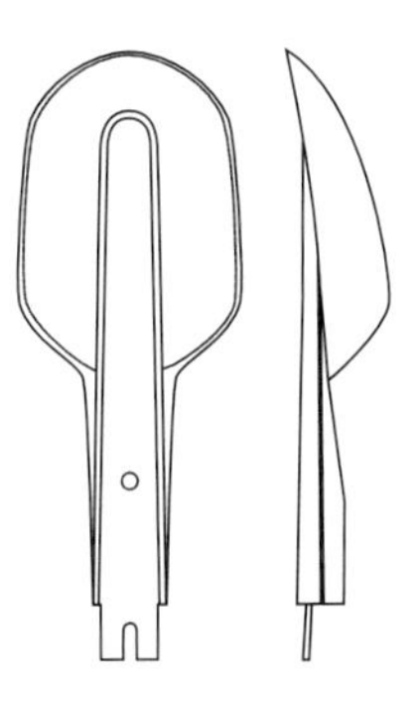

The folded plastic spoon you get with your cookies and coffee on short-haul flights is not a genuine collapsible. It is designed to be used only once.

The fact that a product can repeatedly fold and unfold, collapse and expand, does not of itself make it a genuine collapsible. A chocolate box with a lid, for example, does not qualify. The lid's purpose is both to protect and reveal the contents, so open and closed states alike are active. Unlike a genuine collapsible, it does not fold to save space. The chocolate box is only a quasi-collapsible.

Numerous types of luggage and furniture inhabit the same category as boxes. Their primary function when folded is to protect whatever is stored within. These are only quasi-collapsibles.

Height-adjustable office chairs, barber's chairs and adjustable wrenches are also quasi-collapsibles: they have any number of different active states in which they are more or less collapsed, but they have no passive state. Folding for space saving is not part of their normal function.

A genuine collapsible has one folded passive state and one or more unfolded active states; one space-saving state when not in use and one or more expanded states when in use. If the folded state is both active and passive – think of a pair of scissors – then the object is not a genuine collapsible, but it comes closer to true collapsibility than lidded boxes or height-adjustable chairs, which have active states only. When not in use, scissors are (as a rule) folded. When in use they alternate between folded and unfolded. You cannot use the scissors without folding them.

Degrees of collapsibility
– Non
Self-assembly furniture:
It is only 'unfolded' once
– Quasi
Box with lid:
Two active states
No passive state
– Quasi
Office chair:
Many active states
No passive state
– Quasi
Scissors:
Many active states
One active state doubles as
a passive state
– Genuine
Umbrella:
One active state
One passive state
– Genuine
Foldable, adjustable chair:
Many active states
One passive state

The Ouverture stereo set slides open its glass doors at a mere wave of your commanding hand. It is a quasi-collapsible with two active states – one to allow access and one to protect – but no passive state. Designed in 1995 by David Lewis for Bang & Olufsen.

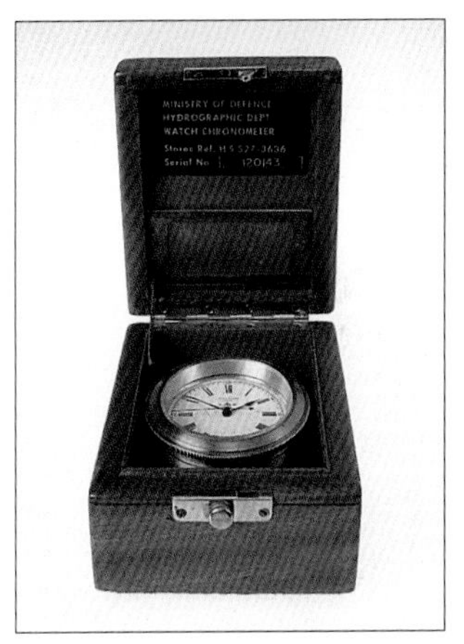

Old British Royal Navy-issue deck watch manufactured by Ulysse Nardin in Switzerland. It is a quasi-collapsible. It has two active states but no passive state.

Height-adjustable office chair. This is a quasi-collapsible with many active states and no passive state. Adjusting it might make you more comfortable, but space-gain is not the point. Designed by Jørgen Rasmussen in 1980 for Fritz Hansen, Denmark.

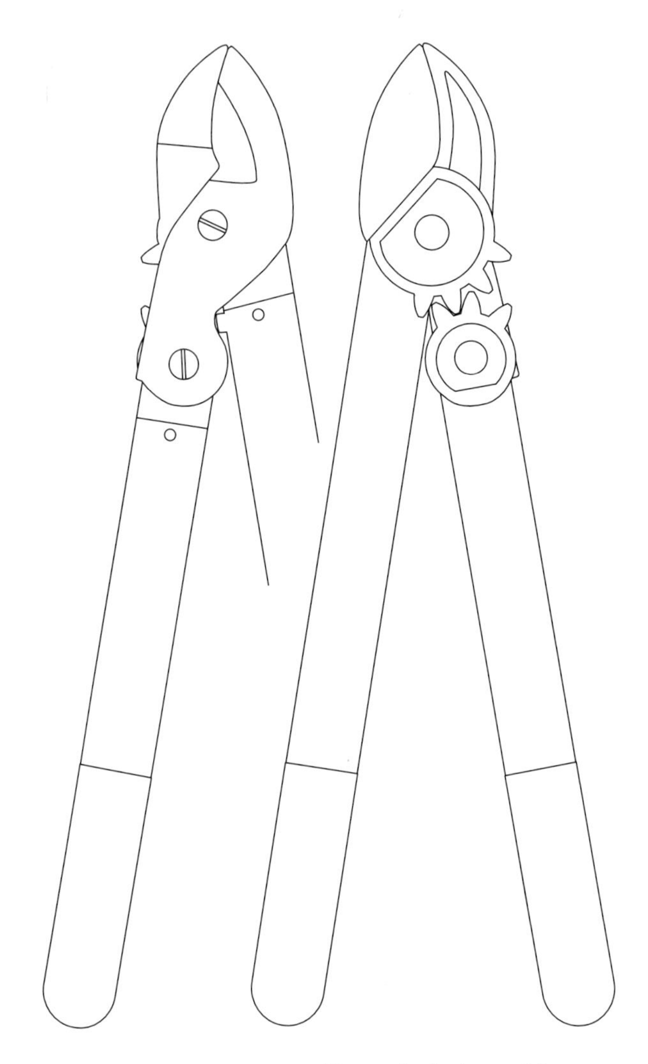

Like a pair of scissors, this compound-action gardener's lopper is a quasi-collapsible with many active states, one of which doubles as a passive state. Cogs on the handles compound the leverage like the gears on a bicycle. It was designed by Olavi Lindén in 1996 for Fiskars, Finland.

Civilizing the outdoors: the great British picnic has inspired many an ingenious collapsible. This cased drinking glass holds a hinged fork, knife and spoon set. The spoon also has a corkscrew that folds out from its handle. The case is a quasi-collapsible, with two active states and no passive state. The fork, knife and spoon are genuine collapsibles.

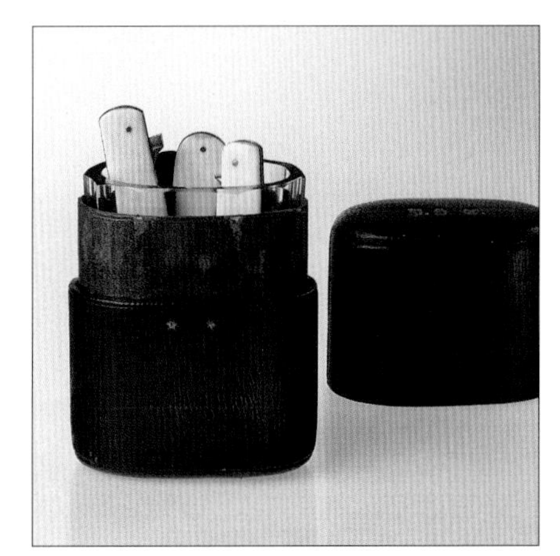

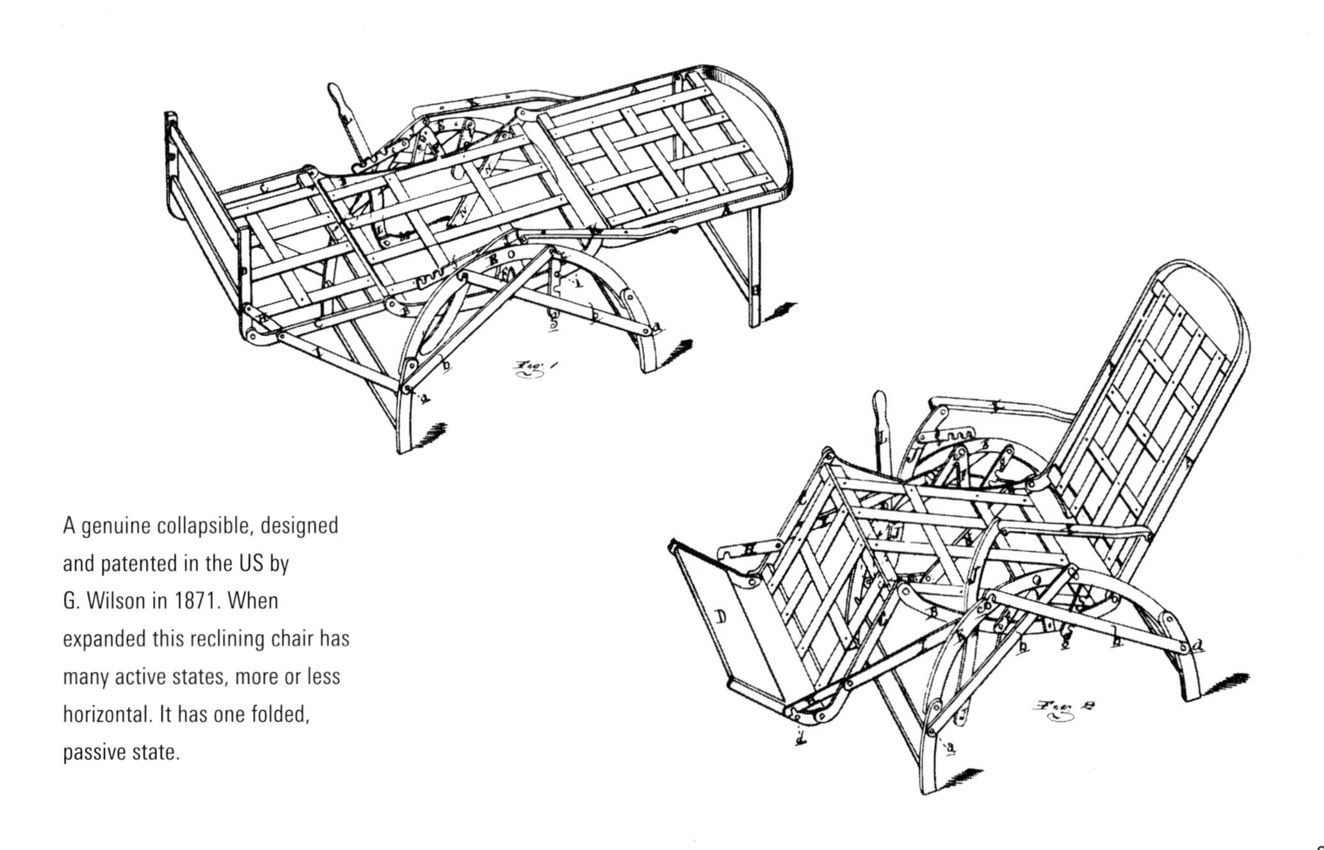

A genuine collapsible, designed and patented in the US by G. Wilson in 1871. When expanded this reclining chair has many active states, more or less horizontal. It has one folded, passive state.

2. Mechanics

Classifying collapsibility

Classification is a useful way of organizing any body of knowledge: studying likenesses and differences between related phenomena always enhances understanding of their nature. Furthermore, classification encourages structured thinking and communication.

Collapsibles, however, defy – or at least resist – a completely rigid classification. The differences between collapsibility principles are often indistinct. The boundary between collapsing by folding and collapsing by creasing (along prefolded lines), for instance, is not always clear. But common sense and a spirit of compromise have prevailed. Between them the twelve categories listed here include all of the most frequently applied methods of mechanical size reduction.

Many collapsibles apply more than one of these principles. When you take down a tent, for example, you dismantle the poles, fold the cover and roll the ground sheet. Those products that apply multiple collapsibility principles are generally shown as examples of just one – probably the dominating, sometimes the most interesting – of those principles.

Twelve collapsibility principles
– Stress
– Folding
– Creasing
– Bellows
– Assembling
– Hinging
– Rolling
– Sliding
– Nesting
– Inflation
– Fanning
– Concertina

These principles are not a set of rigidly parallel definitions. Most describe the action by which an object is collapsed: *creasing*, *nesting*, and so on. Some describe the action by which it is expanded: *assembling*, *fanning*, and so on. *Sliding* describes the action necessary for both. And two – *bellows* and *concertina* – describe not the action but the structure of particular categories of collapsibles.

Zusammenlegbares Militärfahrrad, klein zerlegt, bequem auf dem Rücken zu tragen, Soldat ohne abzusteigen schussbereit.

Bicyclette militaire démontable et repliable, pouvant se porter commodément sur le dos, permet de faire feu sans descendre de l'appareil.

Collapsible military cycle. Easy to carry on back when in pieces. Soldier ready for shot, without alighting.

Bicicleta militar desmontable pudiéndore llevar cómodamente sobre la espalda. Permite tirar sin bajar del aparato.

War generates all kinds of new needs, which have inspired designers and inventors to come up with all kinds of new collapsibles. This hinged German military bicycle from a 1911 Alfa catalogue allowed the biking Fritz to aim, fire and escape without dismounting. When not in use, the bike was easily folded up and carried.

The mahogany room divider PK111, designed by Poul Kjærholm in 1956, dismantles in a matter of seconds into a stack of uniform, easily stored elements. Its volume does not change, but it occupies much less practical space. Manufactured by PP Møbler, Denmark.

Stress

The rubber band works by stress-tension. It expands to grip the produce, then contracts again ready for a new expansion.

The first and in a sense the most basic collapsibility principle is really a double principle. Stress implies both compression and expansion, pressure and tension. When a collapsible works by stress-pressure, its stressed (compressed) state is for storage and its relaxed state is for action. Thus a sleeping bag is compressed for storage and transport, and then released at bedtime. When a collapsible works by stress-tension, the reverse is true: its stressed (stretched) state is for action while its relaxed state is for storage. Thus a stretched elastic strap keeps your luggage in place on the roof of the car, but then relaxes and releases when its fastener is undone.

Both compression and stretching entail an element of violence against more or less defenceless tools. Although they are mechanical opposites, compression and stretching are sometimes practised together. Picture the stressed traveller sitting on his suitcase, struggling to squeeze its contents down to a manageable volume, while simultaneously stretching the suitcase lid to meet the base. The content is compressed. The suitcase is stretched. Man and machine are stressed.

This example illustrates the fact that stress – whether compression or stretching – is often an unofficial collapsibility principle, applied to tools rather more often than their designers might have envisaged.

Backpackers wage a never-ending struggle to minimize the weight and volume of their gear. One way of reducing volume is to stow a sleeping bag in a compression sack, which works by stress-pressure. Manufactured by Yeti, Germany.

The Term-A-Rest traveller's mattress reduces for storage when the air is pressed out and held out by a stopper. When the stopper is removed, the mattress automatically takes in air and reshapes itself. The principle is stress-pressure. Manufactured by Cascade Designs, US.

The photographer's collapsible light reflector was invented *c*. 1985 by photographer John Ritson, who had the idea when he watched a carpenter fold the band from a bandsaw to hang it on the wall. It consists of a spring suspending a reflective sheet that is white on one side, silver on the other. It works by the principle of stress-pressure. When twisted, the spring coils itself into three smaller rings, making it compact enough to stow away in a small bag. The photo shows the reflector half-way between extended and folded. Manufactured by Lastolite, US.

Anomaly

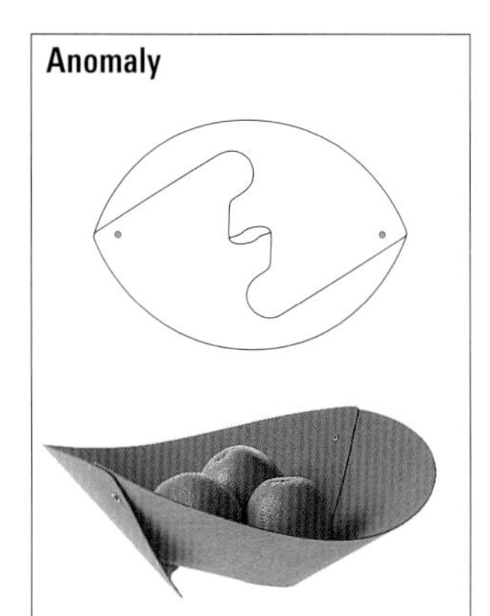

This flexible wooden bowl is an anomaly among collapsibles: it 'folds' to expand and 'unfolds' for reduction and storage.

Folding

One of the commonest types of collapsibility is made possible by the directionless flexibility of soft materials such as cloth and certain types of plastic. This is the quality that allows clothes, towels and blankets to be folded when not in use, curtains to be drawn, or a tent packed away. When it comes to collapsibility, softness can be a strength. It enables us to fold our handkerchief however we like: by careful halving until it achieves a pocket-friendly format, or – for the less particular – by just wrinkling it up into a ball.

Packing clothes for travelling without creasing them is a discipline once perfected by the British gentleman's gentleman, and now is practised by smarter travellers everywhere. Flags, sails and fishermen's nets also fold away for storage.

Paper, however, does not fall under this collapsibility principle. Although it folds, it does not possess the directionless pliability of, for example, cloth. Paper frequently does have direction, and it is less soft than (unstarched) cloth. As a result, folding tends to leave creases which encourage repeated folding along the same lines. And that is the definition of the next principle of collapsibility (see p. 52).

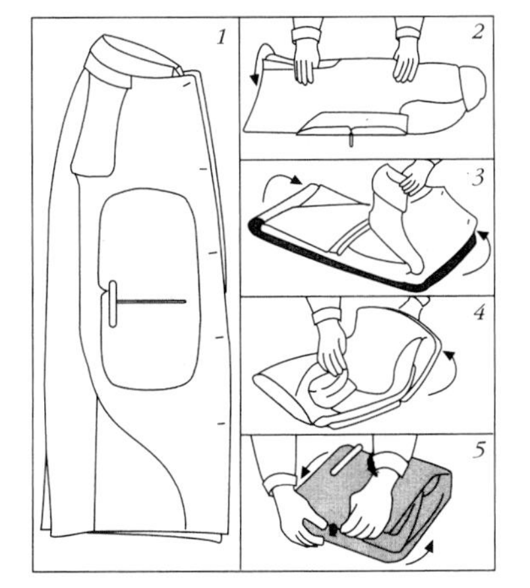

Packing your Acquascutum Voyager raincoat may take a minute or two, but unpacking is done in seconds. Just remember not to fold it wet and not to catch the cloth in the zip.

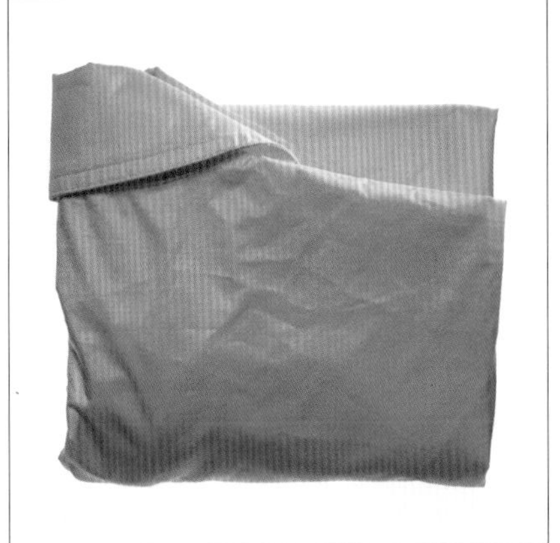

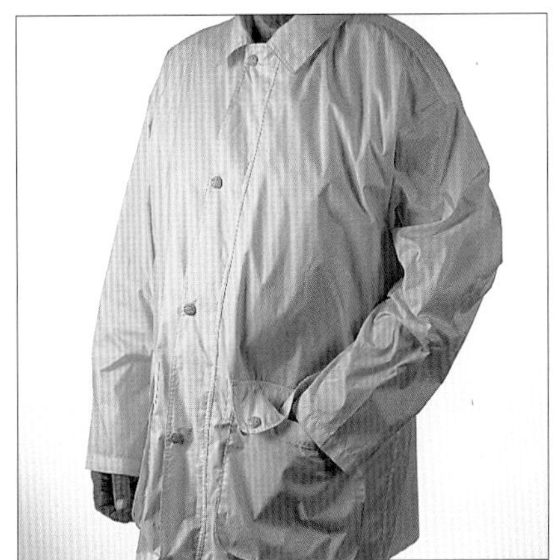

When folded away, this wet-weather jacket by Ermengildo Zegna of Italy is spatially as modest as it is visually conspicuous when folded out and flashed by its owner.

Don't forget your waterproofs at the open-air opera: polythene hoods and capes are widely available, and will save you having to resort to plastic sacks.

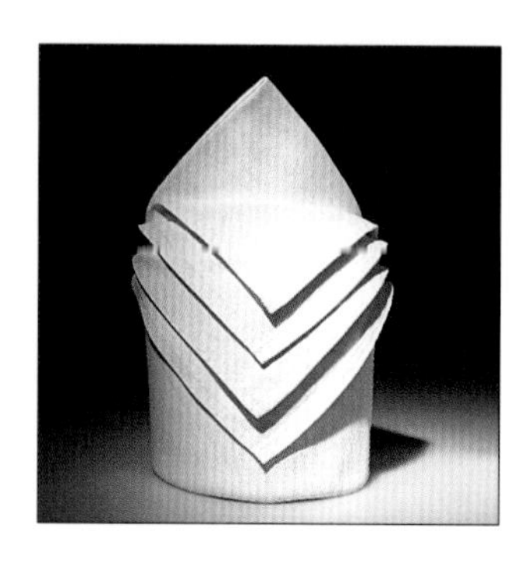

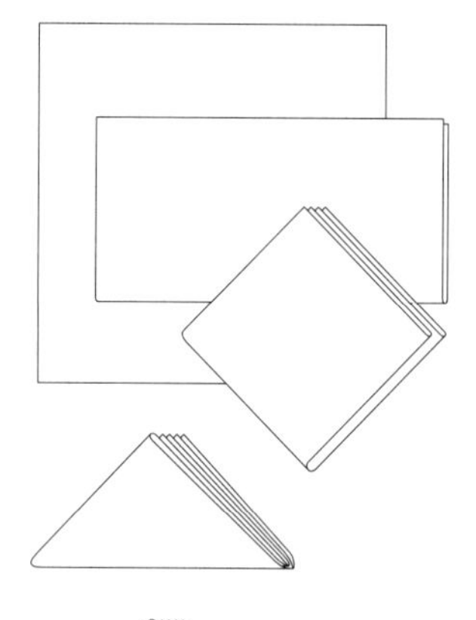

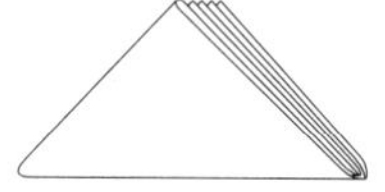

Napkin folding is nothing but a passion for many hostesses. They study books of exotic patterns and rehearse in good time before the party.

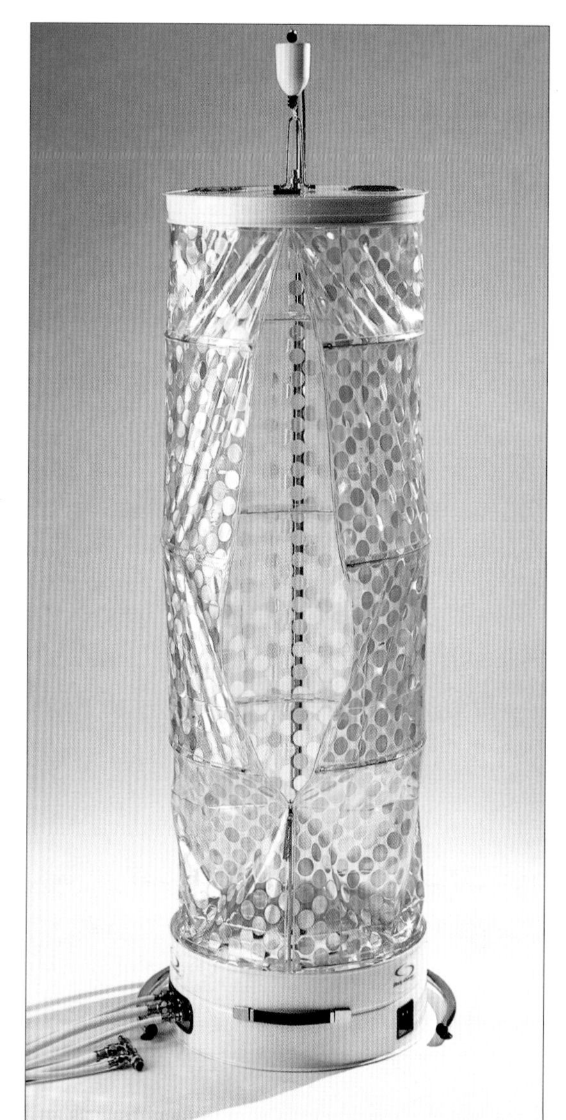

Portable washing facilities: this collapsible shower cubicle connects via a hose to the tap of a hand basin, and delivers waste water back to the basin by a pump at the base. It was designed in 1999 by Michael Perthu and manufactured by Body Shower Mobile, Denmark.

The Bantry Bay Longboat is a reconstruction of a quarterboat attached to an 18th-century French man-of-war that was stranded in Bantry Bay, Ireland. It is fitted out with foresail, mainsail and spanker. All sailing boats are collapsible by folding insofar as the sails can be let out and taken in to meet changes in the weather. The Bantry Bay Longboat, however, is more collapsible than most. Its masts can be dismantled altogether to turn it from a sailing boat into a rowing boat manned by thirteen able men: one skipper, ten rowers and two extras.

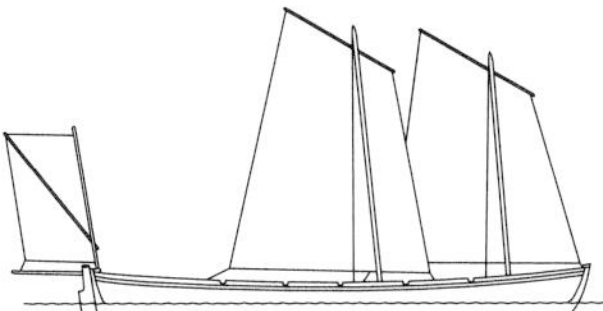

◀

A flag is most beautiful in a head wind goes an old saying. On 20 June 1969 – as on other days on the moon – there was no such wind. When telecast to earth, the Stars and Stripes needed support to assert its message. The softness was partly neutralized.

Many countries have rules for folding their vexillological prides.

The American Stars and Stripes is folded down to a triangle of blue and white.

The Danish Dannebrog is often folded and rolled to hide and protect its white cross.

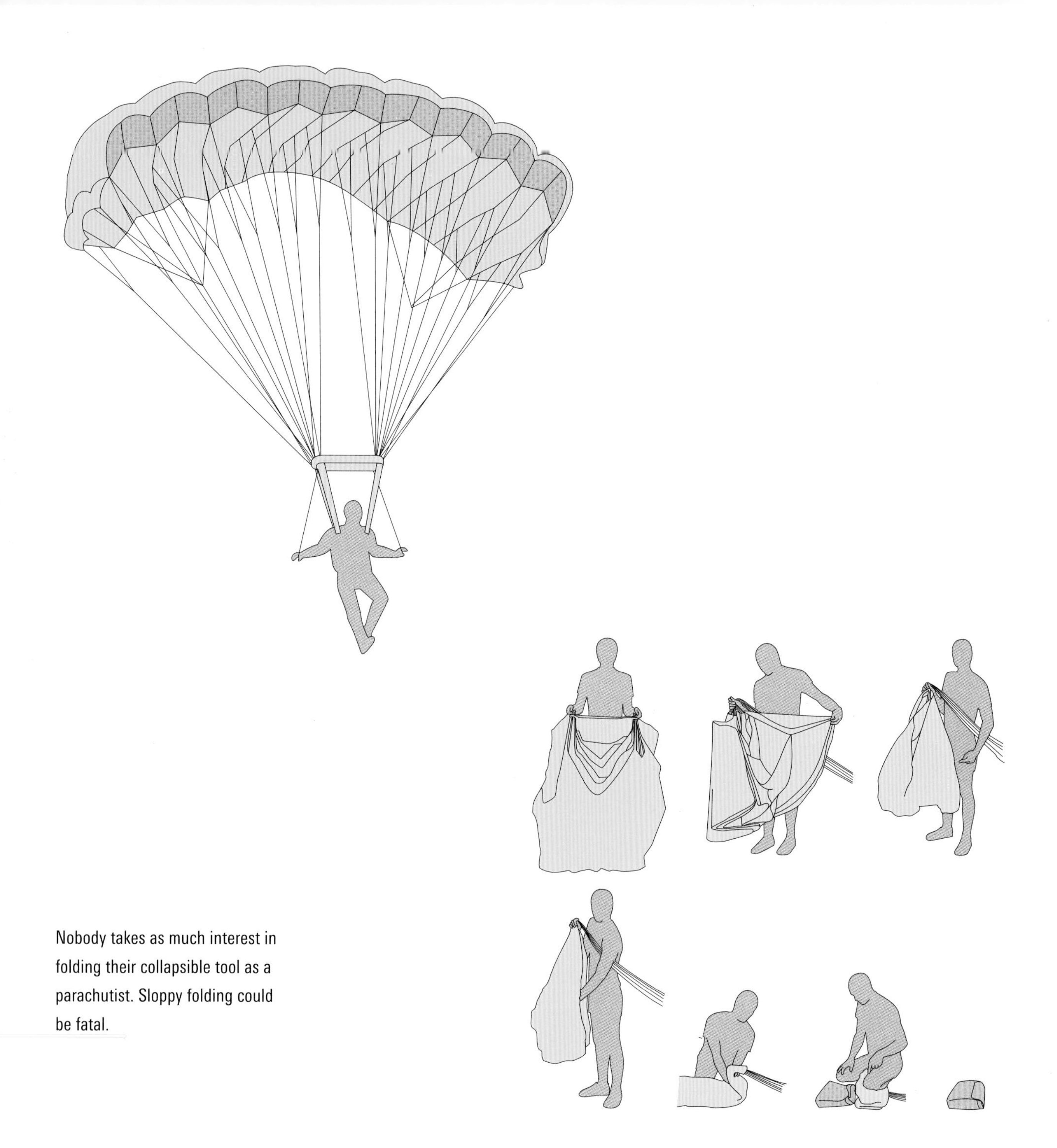

Nobody takes as much interest in folding their collapsible tool as a parachutist. Sloppy folding could be fatal.

Portable transportable: folding pail sold through the American C.M. Moseman & Brother catalogue in the 1890s.

Water sack by Fjällräven, Sweden.

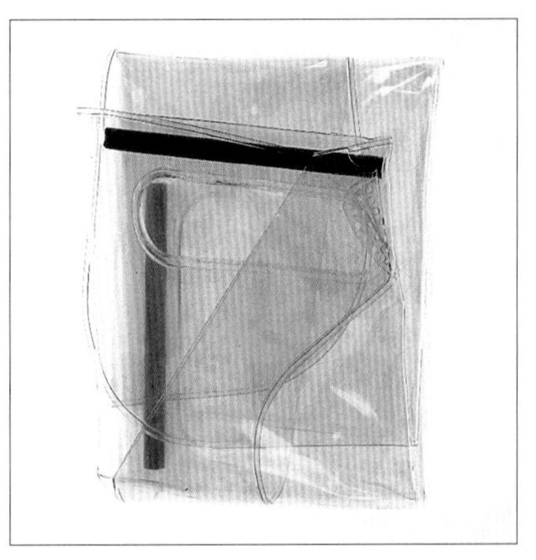

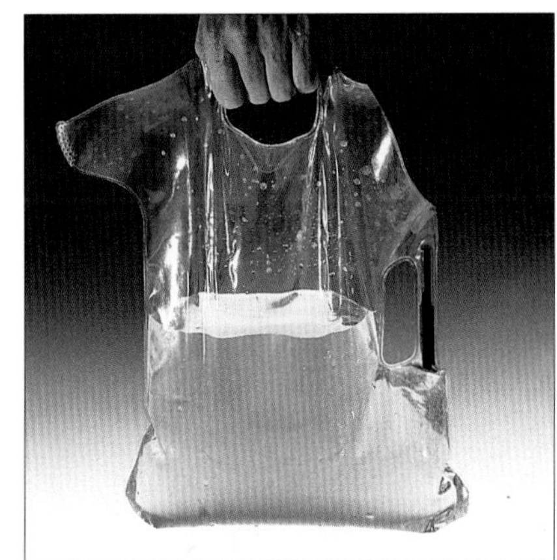

Foldaway plastic watering can by Hagn & Kubala, Austria.

The tote bag occupies no space at all when folded but holds groceries galore when opened out for use.

Any more minimal and the string bag would not be a bag.

Plastic carrier bags are reusable time and again, and qualify as true collapsibles.

Three degrees of protection: the Eureka Hunting and Blizzard Cap, advertised in the 1908 Sears, Roebuck & Co. catalogue, folds out twice for extra warmth.

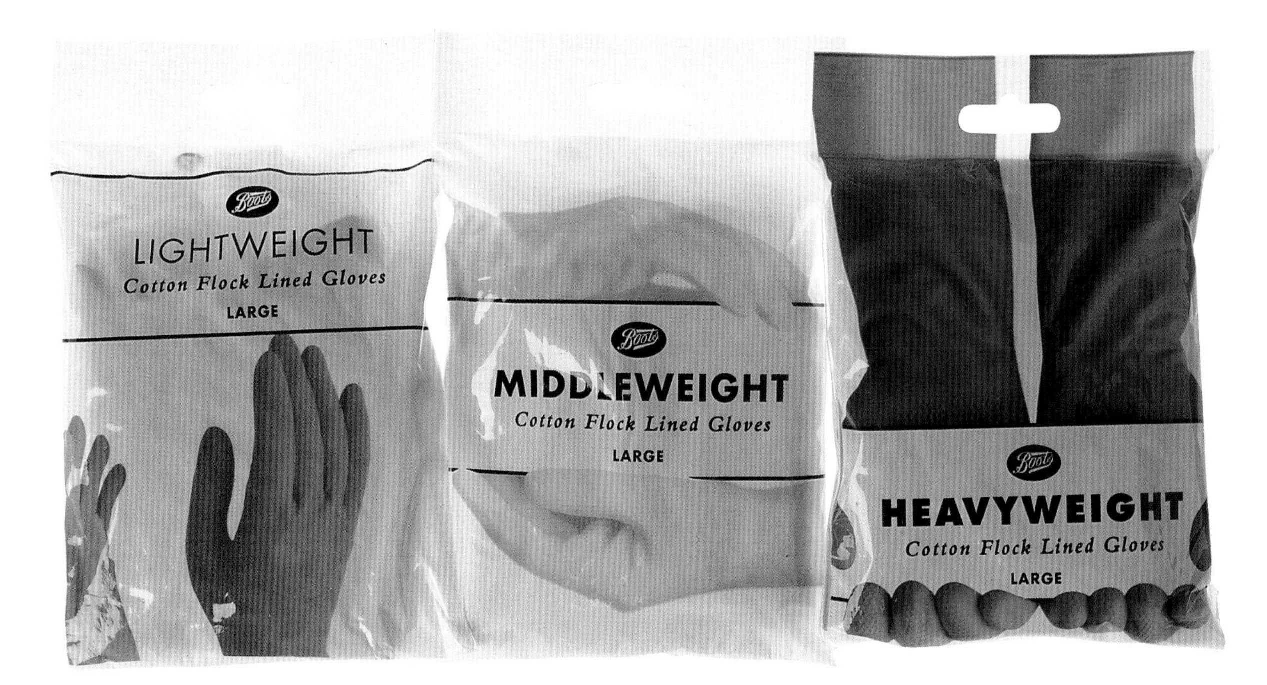

Doctors, policemen and cleaning ladies alike all protect their hands by the same artificial skin layer: latex rubber gloves. They were invented in 1950 in Britain by the J. Allen Rubber Company, which marketed the new product as Marigold Gloves. Today, innumerable makes and varieties are available.

Collapsibility may be one of the reasons for the enduring popularity of the skipping rope. (Compare the short-lived fad for the hula hoop, a toy that would hardly fit in the average school bag.) To reduce friction, skipping ropes have been made with ballbearings in the handles. However, the version with simple hollow wooden handles remains the definitive form. The origins of the skipping rope are unknown but, boxers aside, this genderless toy has inexplicably become a symbol of femininity.

Bedouin black tent, southern Morocco. Tents are home to man on the move. Nomads populate the marginal parts of the habitable world – desert, steppe and tundra – and move to find food for themselves and their animals. They have developed tents of many kinds. While vernacular architecture is traditionally determined by two factors – the materials at hand and the purpose in mind – the tents of nomads are additionally constrained by the load capacity of their transport animals. The tent of the Bedouin, nomad of nomads, is designed to be transported on camel back. Woven from the hair of black goats, its purpose is to protect a family against sun, cold, wind, sand and dust – and to provide privacy. The roof is made of long widths of cloth stitched together. Across these are sewn tension bands to which long stays are fastened. The bands take the strain from the stays. The stays may be very long to absorb wind shocks, to give perpendicular pull from the stakes and to discourage unwelcome visitors. The roof is supported by nine or more wooden poles. The walls are pinned to the tent roof. Inside, a curtain divides the men's area from the (larger) women's area.

►

Recreational nomads in the West constitute a market for tents ranging from minimal survival shelters to luxury textile villas. The traditional triangular tent construction is based on one or more poles on which are concentrated the weight of the roof and the tension created by the guy lines. The poles and the roof are mutually dependant; neither can stand alone. The triangular design is typified by these tents from the 1908 Sears, Roebuck & Co. catalogue. This type of tent is still available today, but more up-to-date designs are often self-supporting.

:PARTMENT = OF = TENTS, WAGON, STACK AND MACHINE COVERS AND AWNINGS

: OWN AND CONTROL ONE OF THE LARGEST EXCLUSIVE TENT, FLAG AND AWNING FACTORIES IN THE UNITED STATES. WE ARE ABLE TO RUN THIS FACTORY : YEAR AROUND MANUFACTURING THESE GOODS, WHICH WE OFFER YOU, QUALITY CONSIDERED, AT PRICES FROM 25 TO 50 PER CENT LOWER THAN YOU :LD PURCHASE THE SAME GRADE OF GOODS ELSEWHERE. OUR TENT, COVER AND AWNING BUSINESS IS THE LARGEST IN THE UNITED STATES TODAY OF ANY HOUSE DEALING DIRECT WITH THE CONSUMER.

ME YEARS AGO we were buying tents and covers the same as other dealers are buying them today—in the open market; but we found : we could not, as our business increased, give our customers goods which we could :ly guarantee at all times to be up to grade. There is no business today that is so de- :alized as the tent and awning business. Nine out of ten dealers today sacrifice quality in : goods in order to offer competitive and attractive prices. It was this condition that :ically forced us into the tent manufacturing business. When we found that we could safely recommend the tents and covers we were buying in the open market, when we :oughly realized the conditions under which these goods were manufactured and sold :ughout the country, we resolved that either we would sell honest tents and :st covers, or we would quit the business, and we found that the only way in which :ould sell honest goods in this line, goods that we could absolutely guarantee to be full :and full weight, was to manufacture them ourselves. We accordingly started a factory :r own, and in the short time of three years this factory has grown to be one of the :st factories of its kind in the United States. Our growth in this business is not due to the that we have sold certain size tents and covers for less money than other houses, but is entirely to the fact that the tents, covers, awnings and flags we sell are exactly as we :esent them to be in every particular, and that when we state a tent is 9½ feet wide by :et long, the tent will be found exactly in accordance with these measurements, and not :hes short in the width and 5 inches in the length, height of wall or height of center.

E AVERAGE MAN who buys a tent or cover knows little about what he is buying. He is apt to assume that a tent is a tent and a cover is :ver, and in making his selection will, naturally, be guided largely by the price. Lack :pace will not permit us to explain in full in how many ways a tent is "skinned" by un- :pulous dealers and manufacturers in order to be sold at or below our price. For your :ance we will state a few methods employed by most of the manufacturers and dealers :y: The average 8-ounce tent sold by other dealers is made of 7-ounce duck; 9-ounce : is invariably substituted for 10-ounce. A tent properly made, such as ours, ought to :ade of 29-inch duck; this is the standard width, but in order to effect a saving, many :ufacturers, in order to substantiate their claim that their tents are made of 8-ounce duck, :uck 36 inches wide. The wider the duck, naturally the lighter the material. Eight-ounce : 36 inches wide is not as good a piece of material as 8-ounce duck 29 inches wide, such :e use. A common plan of deception is to make the door of the tent of lighter canvas than :body of the tent. All our tents are made with doors from 15 to 30 inches wide, in pro- :on to the size of the tent, and you will find the same material in the door of the tent as :will find in the roof or body. Jute rope is also used very extensively in competitive tents. : rope is entirely unfit for tents, as it rots very quickly and becomes hard and stiff, and :e is always the danger of having your tent blown down at the risk of considerable incon- :ence and possibly danger, when jute rope is used. We use the very best pure sisal and :ila rope, and in all tents over 12x18 feet, and in all refreshment and photographers' tents :se the best pure manila rope.

: WANT YOU to know all about our tents, covers and awnings. The more you know about them, the more certain we are of receiving your :r. To aid you to buy to your best interest, we will send you upon request a free sample : of canvas, showing the quality of duck we use in our tents and covers, also giving on :everse side of the card instructions on "How to judge a tent."

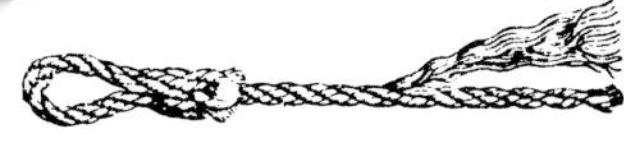

Fulton Brands.

Other Makes.

:IS ILLUSTRATION will give you some idea of the relative qualities of jute rope, used by other manufacturers, and pure manila and sisal, used on all our Fulton Tents and Covers. Note the short, scanty fiber in the jute rope, indicating that the rope has absolutely no body to it, consequently no strength. Also note the common wire splice used by almost all other manufacturers. Compare this with the rope we use, as shown in the illustration, and note the long, full fiber stock, giving this rope enormous strength, and also note the hand worked sailor splice we use as compared with the wire splice used by others.

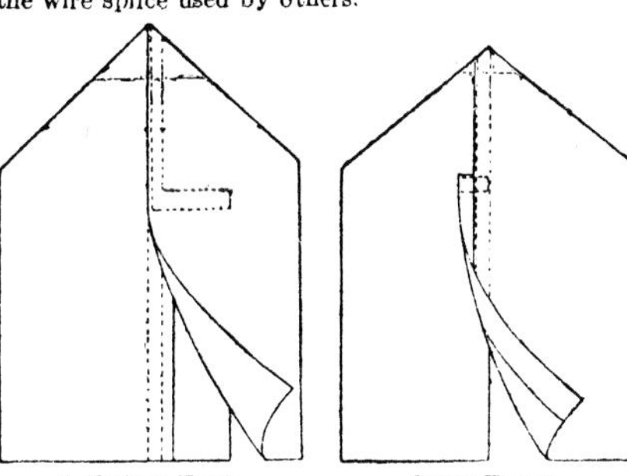

A Fulton Tent. Other Tents.

THIS ILLUSTRATION shows the difference between a Fulton Tent as compared with any other make. Note the large, full door flap used by us, and note that our door laps over almost a full width of canvas, whereas the doors offered on other tents lap over but a few inches. We could reduce the price of our tents materially by making a cheap door, the same as that shown in the competitive tent, but in our Fulton Tents and Covers, we aim to represent only the highest quality that can be made in this line. Our object is not to see how cheap we can furnish these goods, but rather how good we can make them and then make the price consistent with the high class of material, labor and workmanship employed.

TWO DOORS. On all wall tents that are 16x24 feet or larger we will furnish a door at both ends without additional charge. We believe that a tent 16x24 feet or larger ought to have a door at both ends on account of its size, as a matter of convenience and to secure proper ventilation. We, therefore, will furnish all tents 16x24 feet with a door at each end, if desired, at the regular price quoted. Be sure to state whether you want two doors in your tent if you order a tent 16x24 feet or larger. On tents smaller than 16x24 feet we will furnish an extra door for 60 cents.

PRECAUTION. Be sure to slacken your guy ropes on your tent after or immediately before a rain storm, otherwise when your tent dries, the canvas will naturally shrink somewhat during the drying, and your tight guy ropes will pull the tent out of shape. Slacken the guy ropes until the tent is thoroughly dry, then again make them taut.

QUALITY AND PRICE GUARANTEE. We guarantee every tent in this catalogue to be lower in price than the same quality of goods can be bought elsewhere. If any house ever meets or cuts our price on any article, they do it at the expense of quality. If you do not find this so by comparing the goods, or if you ever buy anything from us that is not lower in price than the same high quality of goods can be bought from any other house, you are especially requested to return our goods at our expense and get your money back at once. We guarantee our tents and covers to be exactly as described in our catalogue. We guarantee them to be full weight 29-inch canvas, and to be full size. We guarantee the door of the tent to be of the same weight canvas as the rest of the tent. If you purchase a tent of us and do not find it exactly as we represent it, you have the privilege of returning it to us at our expense and your money and transportation charges will be refunded to you. Weigh one of our tents or covers and one of our competitors' make and note which weighs the most.

DELIVERIES. We carry all sizes of tents from 7x7 to 12x12 feet in stock and make prompt shipment, and on all other sizes allow three to five days' time to make the tent. In June and July allow from four to eight days.

FULTON WALL TENTS.

HIGHEST QUALITY.

We give weight of tents with poles below on 8-ounce tents. 10-ounce will weigh about one-quarter more and 12-ounce about one-half more than 8-ounce. **The weights may vary slightly, as poles do not always run alike.** A 9½x12-foot tent makes a good outhouse or summer kitchen.

We are the only manufacturers of the celebrated Fulton Wall Tents. We warrant them to be exactly as represented. **In ordering, give catalogue number, length, breadth and price. Allow three to five days' time for making tents, and in June and July allow four to eight days, according to the number of orders we have on hand at the time we receive your order. Send for samples of canvas which goes into our tents and instructions on "How to Judge a Tent."**

We carry all sizes of tents from 7x7 feet to 12x12 feet in stock and make prompt shipment, and on all other sizes allow three to five days' time to make the tent. In June and July allow from four to eight days.

Warning—We do not use jute rope in our tents. We use the best pure sisal rope up to and including the 12x18-foot tent; larger than 12x18, we use the best pure manila rope. On all wall tents size

No. 6K10350 Wall Tent.

:2 and larger, we furnish top guy ropes at each end. Order by catalogue number and give size of tent and weight of duck wanted.

No. 6K10350

:gth and Breadth	Height of Wall	Height of Pole	Weight, 8-ounce	Our Prices include Poles, Pegs, Guys, Guy Ropes, etc., complete, ready to set up.		
Feet	Feet	Feet	Pounds	8-oz. Duck	10-oz. S. F. Duck	12-oz. S. F. Duck
7 x 7	3	7	30	$ 5.60	$ 6.70	$ 7.90
7 x 9	3	7	35	6.65	7.84	9.40
9 x 9	3	7½	40	8.00	9.54	11.44
9½x12	3	7½	45	9.00	11.00	13.15
9½x14	3	7½	50	10.40	12.50	14.90
12 x12	3½	8	55	11.15	13.20	16.00

"A" or Wedge Tents.

We do not ship Tents C. O. D.

FULTON BRAND

The weight which we give includes poles. When poles are not wanted with tents deduct 5 per cent from the price of 8-ounce tent. Give catalogue number and style wanted.

No. 6K10340 The following prices include poles:

Style No.	Length and Breadth	Weight, 8-oz.	Height	Price, 8-oz. Duck	Price, 10-oz. Duck	12-oz. S. F. Duck
A	7 x 7 ft.	25 lbs.	7 ft.	$4.59	$5.50	$ 6.40
B	7 x 9 ft.	27 lbs.	7 ft.	5.63	6.73	7.96
C	9 x 9 ft.	32 lbs.	7 ft.	6.31	7.57	9.02
D	9½x12 ft.	38 lbs.	7½ ft.	7.61	9.00	10.92

The Fulton Rope Ridge Wedge Tents.

As more and more of us find ourselves doing sedentary indoor work, we seek to compensate for the lack of exercise by devoting our leisure time to sports and physically demanding pastimes. These out-of-work activities are serious business, and the equipment with which we pursue them can't be too good: we want to be professional even when taking time off from our profession. Advanced gear is a pleasure in itself and it can spare us the more unpleasant aspects of any activity: genuinely waterproof boots, lighter tennis rackets, windproof tents, and so on. Some campers are utterly dedicated to making their villa-tent or autocamper a mobile replica of the home they have left behind: no time without TV and fridge seems to be the rationale. The Scout movement thinks differently, however. This Danish Scout camp is a collapsible theme park of old-fashioned tent technology. Boy Scouts are required to experience the outdoors armed only with survival essentials and minimal comforts. When it comes to struggling with the elements, bad experiences make good memories.

Tents have been around since biblical times, yet new designs are still being developed. The creation of ever more sophisticated materials fuels this progress. Tent designers must accommodate a number of conflicting demands for easy handling, maximum shelter and minimum weight, and they make great efforts to find new spatial and technical solutions to meet these demands. At the same time developers of tent fabric seek to create new materials which let humidity out of but not into the tent. Old-fashioned cotton does that to a degree, but it is heavy and prone to mildew.

American designer Bill Moss has been at the leading edge of tent design in recent years. His 1991 Little Dipper (above), a high-performance three-man tent produced by Edgeworks in the US, incorporates a number of innovative features and is one of his most spectacular designs. Its curved-pole construction is both stable and streamlined. The poles are sleeved which means that the tension is distributed over the entire structure.

Advances in peak performance tents have also spread to backpackers' and family tents, such as Bojum, also by Edgeworks, which houses up to five persons and maybe a little dog.

Tents protect against both the climate and prying eyes. This Artists' Portable Tent from the British Army & Navy Stores catalogue of 1907 served the performers of yesteryear.

Architect Arne Jacobsen's commissions of 1932–34 at the Bellevue Beach north of Copenhagen comprised villas, apartment blocks, restaurant, theatre complex, seaside kiosks – and these fabric changing huts.

Fast construction, high mobility, effective weather protection and low costs have always made tents the shelter of choice for entertainment, celebrations and travelling shows. For the same reasons, tents are the obvious solution to sudden homelessness due to natural disasters or war. Tent-like constructions are also used in many trades to provide temporary protection from sun or rain. The shelters produced by Swedish company Hallbyggarna are widely used in the construction industry to provide protection when casting concrete, for storage and more.

Large-scale tent constructions remind us that tents are subject to the same physical laws as any other construction, animate or inanimate: as they get larger, they need more supporting structure. Industrial tents require bigger, more complex skeletons than a backpacker's canvas.

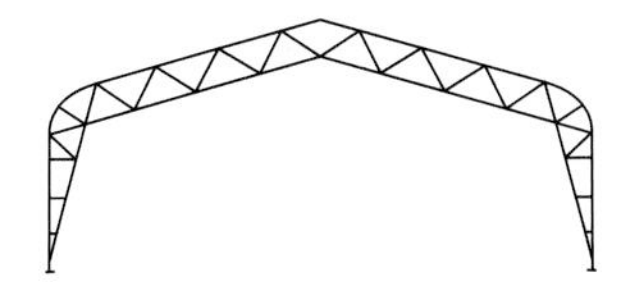

A collapsible tent can proclaim the identity of a business, making it highly visible in the streetscape. The tent can also protect the owner, the merchandise and the clientele against climatic wear and tear. Its design may be anything from a simple suspended sail keeping off the sun to complicated, semi-permanent constructions.

A circus is not a circus without its big top. At the Danish circus Dannebrog, the textile collapsibles are all colour coordinated: big top for the performance, t-shirts for the clowns.

Creasing

Vintage French postcard that folds and unfolds along prescored creases to allow stereoscopic peeping.

Instead of being randomly wrinkled, a piece of cloth or other pliable material may be folded along preset lines or creases. The purpose of creases may be two-fold. First, they can give an object, both folded and unfolded, a neater appearance than random folds could ever create. Second, creases may facilitate the acts of folding and unfolding and permit a greater size reduction in the collapsed object.

Maps of many kinds are folded along creases to make them as handy as possible for the traveller. Much creative thinking has been devoted to developing ways to fold a map that provide easy access to every part of it without having to lay out the whole thing.

Creases in paper maps are weak points along which the paper will eventually break. To overcome that flaw, early survey maps were cut into rectangles and mounted on strong cloth to make the folds more durable. Escape maps issued to military pilots operating behind enemy lines are printed on both sides of fine cloth for silence of use, durability and maximum compactness.

Pleats are arrangements of parallel creases; they exist only in the plural. Fashion designers have long used pleated fabric for aesthetic effect and to introduce disguised fullness to women's clothing. Issey Miyake has recently brought the look up to date, and brought himself worldwide success. Folding doors and window screens are another familiar application of pleats.

Cutting edge creases: Rudolph Valentino in 1925, sporting chic sharp tailoring.

The horizontal crease around the circumference of this plastic mug makes it possible to fold the upper part inwards to halve its size when not in use. The handle helps unfold it.

Regulation creases: Royal Navy seamen used to fold the trousers of their shore-going uniform with horizontal creases precisely measured to the width of their pay books. Tall sailors would end up with seven creases, smaller sailors with as few as five.

It is not possible to transform a double-curved three-dimensional surface into a two-dimensional plane without a degree of inaccuracy. If the curved surface is that of the globe and the plane is a map, the inaccuracy means that areas, distances, shapes, angles and directions can never all be correct at the same time. This is why geographers over the centuries have developed a variety of different projections, some aiming to show precise distances, for example, others precise areas or directions. Trying to fold a sphere from a flat piece of paper demonstrates the same problem in reverse.

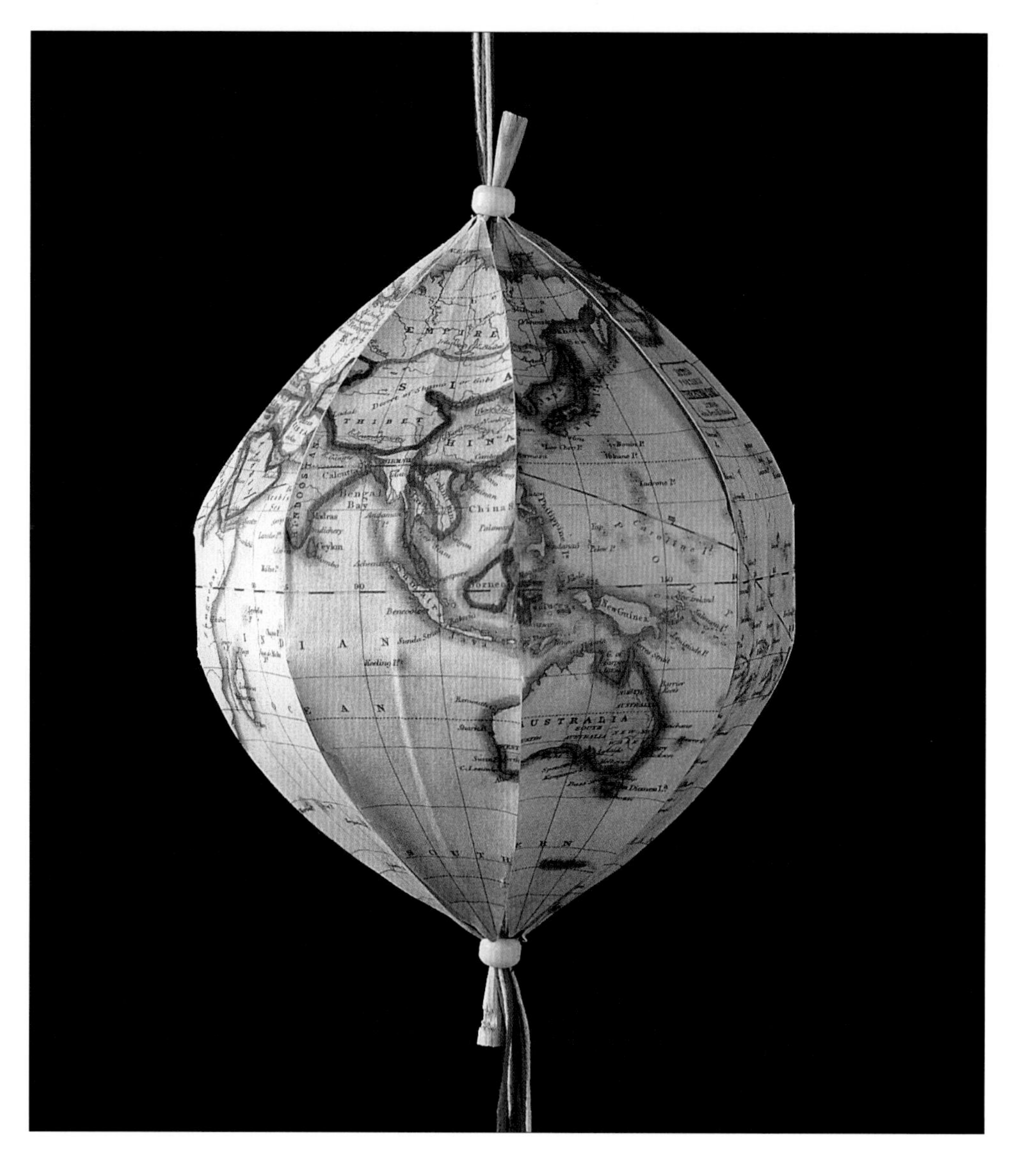

Betts's Portable Globe was devised in the 19th century by the Englishman John Betts. It consists of eight linked leaf-shaped sections. When its strings are pulled together, the tips of the leaves meet to form the poles. The leaves fold along preset creases to meet in a ring around the equator, and the whole forms a three-dimensional sphere-like shape. According to Betts his globe was intended for explorers, the military, the government, students, scholars and missionaries.

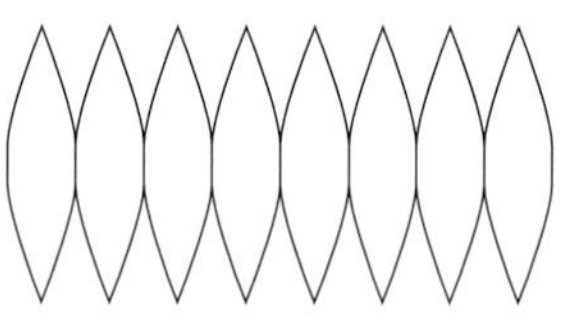

Military escape maps – one of Indochina, the other of Iraq – issued to American and British pilots in case of being shot down behind enemy lines. They are printed on both sides of fine cloth, with preset creases for compact folding.

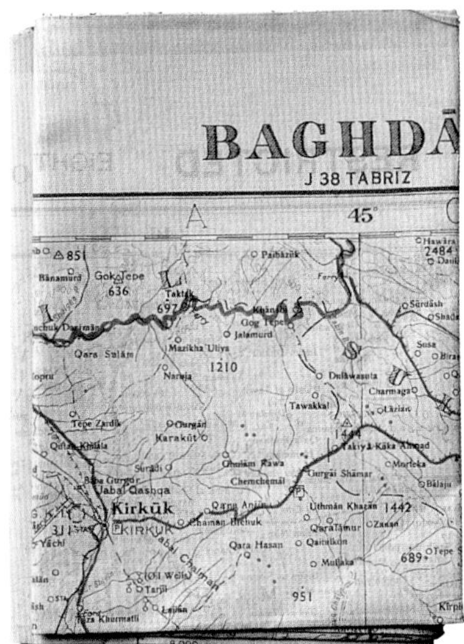

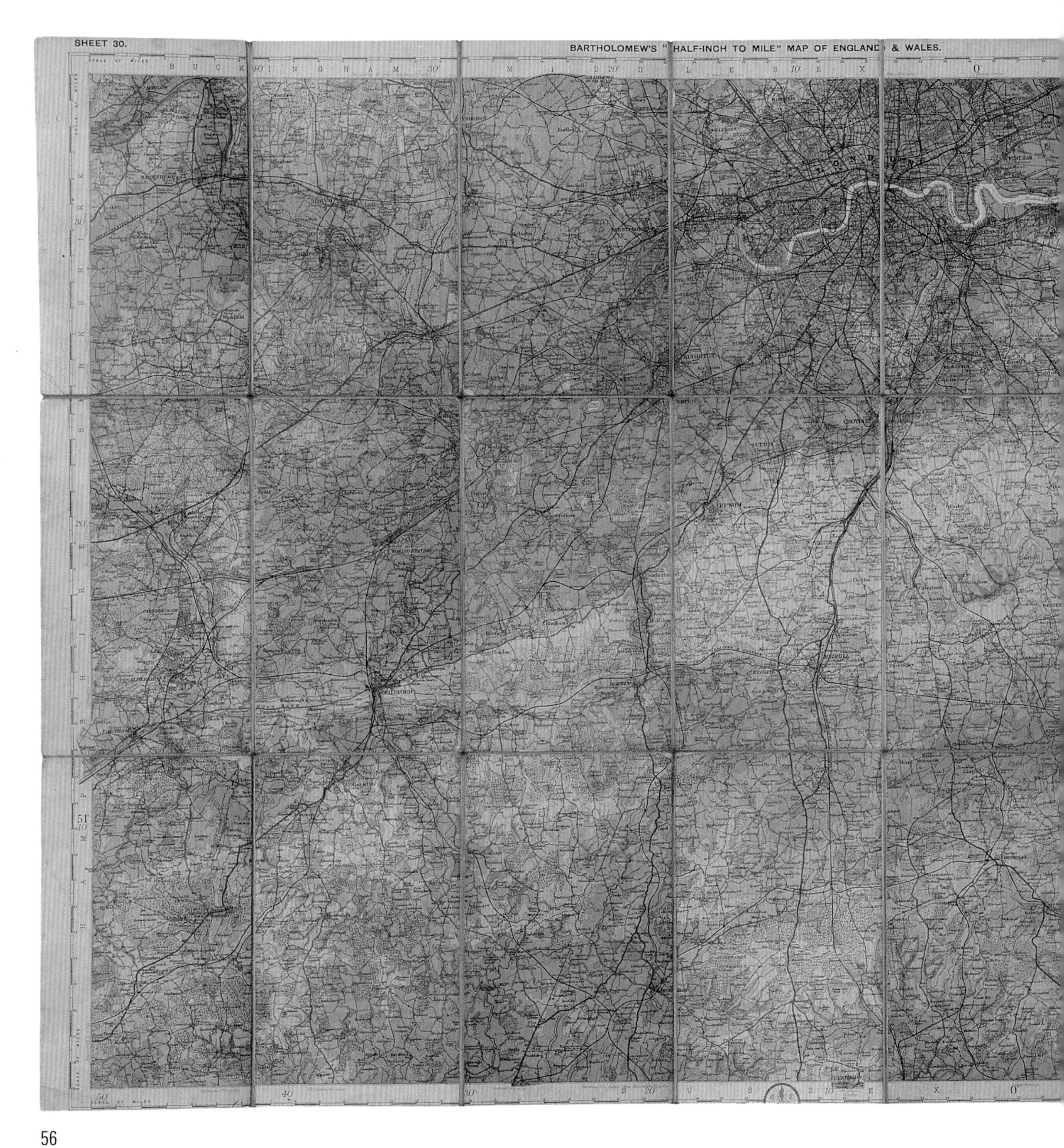
SHEET 30.
BARTHOLOMEW'S "HALF-INCH TO MILE" MAP OF ENGLAND & WALES.
EPSOM
SUTTON

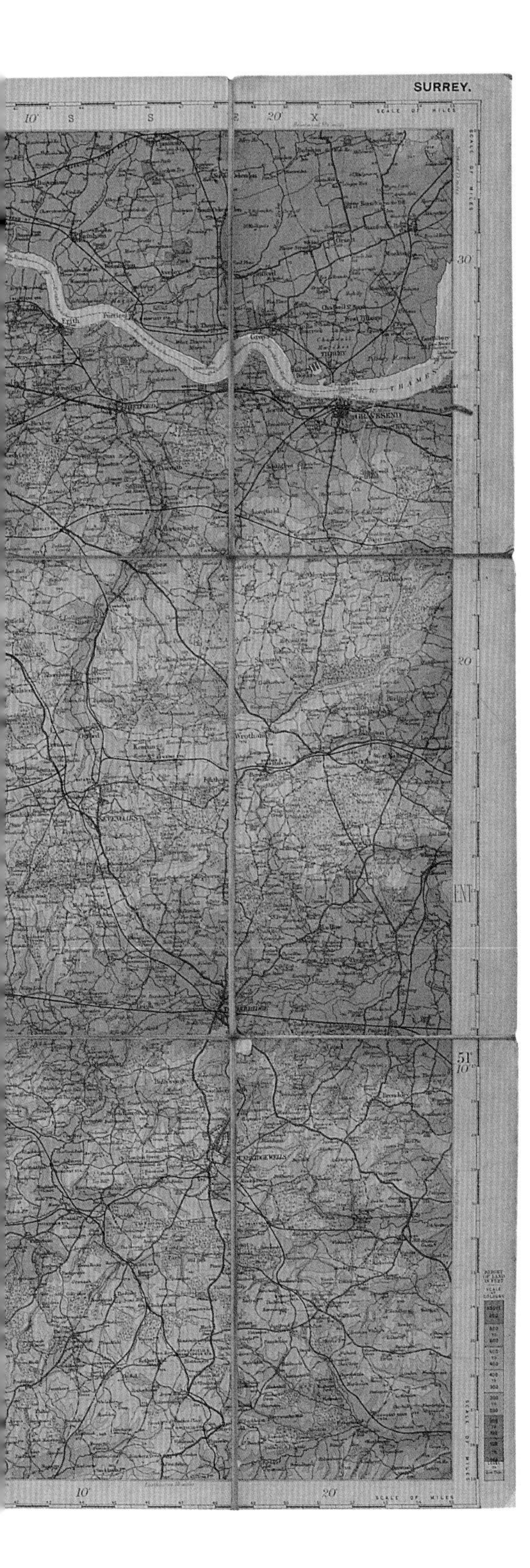

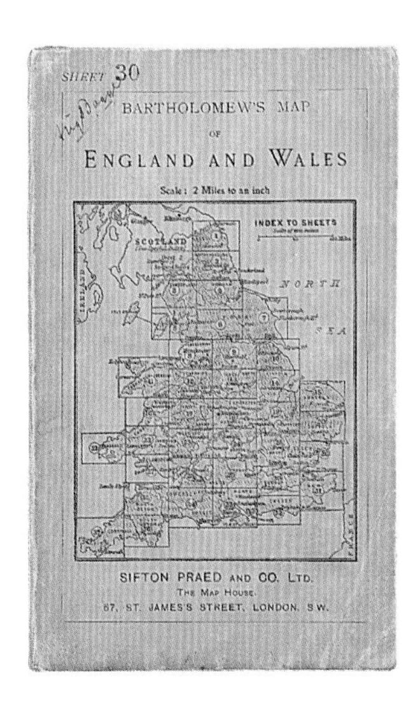

One of a series of old survey maps covering the British Isles, published by Bartholomews. It is mounted on soft but strong cloth for ease of use and longevity.

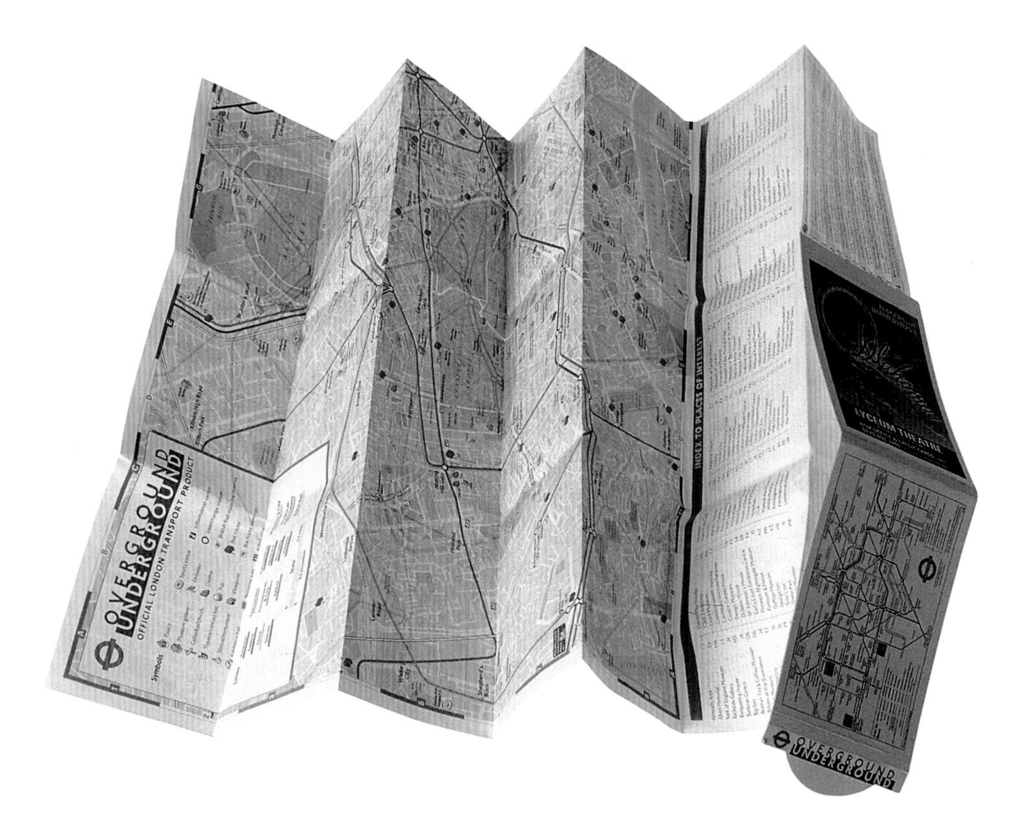

London Overground/Underground map by Snap-Map. The sheet is folded first along a number of horizontal creases, then along vertical creases. Architectural drawings are folded the same way but in reverse – vertical then horizontal. The resulting combination of parallel and cross folds creates a convenient format and allows ready access to any part of the map with minimal unfolding. Thus it is equally possible to gain an overview and to study details.

When you pull apart the front and back of this credit-card-sized Norwegian train timetable, the whole thing unfolds itself.

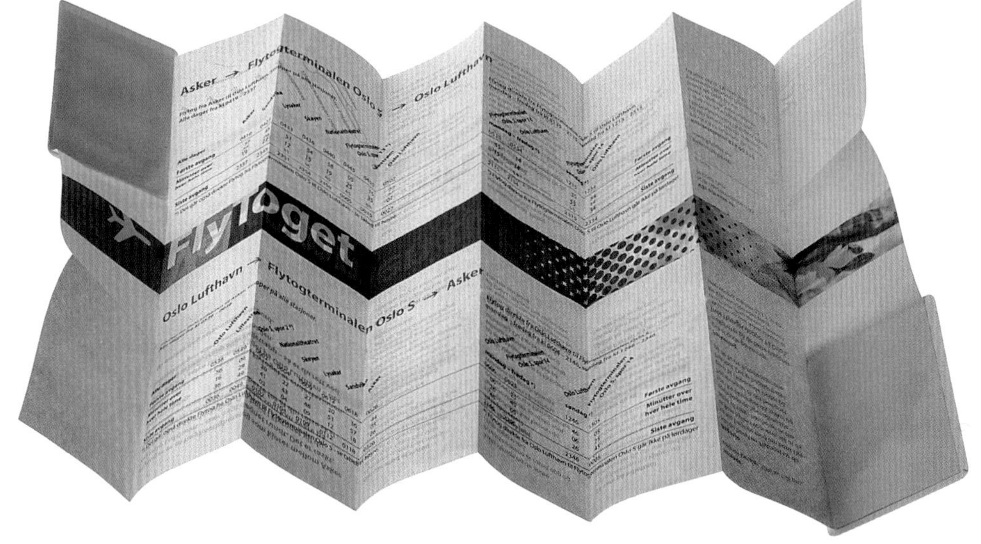

Stockholm folds out for inspection on this map handed out free by the city's cabdrivers. Produced by CR Grafiska.

The VanDam plan of New York City includes three ingenious pop-up maps covering the Big Apple's Up-, Mid- and Downtown. The maps refold unassisted.

A book is a very special collapsible. When closed it has a convenient and portable format. When open, it has as many unfolded states as it has spreads. The predecessors of the book as we know it were creaseless – tables of clay and continuous rolls of papyrus or parchment. The first bound books were made in the early Christian era. Materials and methods of production may have changed, but the book has remained much the same ever since. Folding large sheets of paper is a time-honoured part of the bookbinder's trade. It is a means of reconciling the contradictory desires of printers, who prefer to print on large sheets because it is more economical, and readers, who prefer their books and other printed matter in a manageable format. The bookbinder folds large printed sheets into leaves that are trimmed on three edges, leaving the central crease to open and close like a hinge once it is attached to the spine. These central creases facilitate the repeated transformation between expanded and compacted, which is the defining characteristic of a collapsible. Guidebooks often go further. Many combine a practical pocket-friendly format with fold-out maps, charts and plans. The linen bindings and rounded corners of old-fashioned guidebooks allowed heavy-duty use while remaining easy on the pocket and pleasing to the eye. Many included one or more silk ribbon bookmarks.

THEATRO
DEL MONDO
DI ABRAAMO ORTELIO.
Nel quale distintamente si dimostrano, in Tauole, tutte le Prouincie, Regni, & Paesi del Mondo;
Con la descrittione delle Città, Territorij, Castelli, Monti, Mari, Laghi, & fiumi; Le Popolationi, i costumi, le ricchezza, & ogn'altra particolarità.
Ridotto à intiera perfettione, & in questa piciol forma, per maggior commodità de' Viaggianti.
CON LA TAVOLA
delle cose più degne, che nell'Opera si contengono.
Venetia. Appresso il Turrini.

The plates of this 1603 atlas by Turrini measure just 75 x 105 mm (3 x 4⅛ in). Such miniature atlases were the first books produced in 'pocket' format to save money and space. They were *epitomes* or copies, the earliest issued a few years after Dutch cartographer Abraham Ortelius published the world's first atlas, *Teatro del Mundo*, in 1574.

The German publisher Karl Baedeker founded his publishing house specializing in travel guides in 1827. Baedeker guides were the first to combine sound practical advice with scholarly accounts of a region's art and architecture. They were standard equipment for generations of knowledge-hungry tourists who wanted to know more than they could see for themselves. Some of the guides have beautiful fold-outs, such as this view of the Mount Blanc chain seen from Flégère, from the *Baedeker Schweiz* of 1927.

In opera-loving Denmark, a zig-zag folded piece of paper is called a *leporello,* after Don Giovanni's footman in Mozart's opera of the same name. Leporello carries a folded list of his master's female acquaintances. Convenient format and great access: a *leporello* is ideal for small maps and other tourist literature.

The French *Guides Michelin* are the ultimate arbiters of taste for the epicurian tourist. One or more stars bestowed by anonymous Michelin inspectors can turn an unknown eatery into a roadside sanctuary. Conversely, the loss of stars has allegedly driven shamed chefs to suicide. The graphic symbols used in the guides are themselves semantic collapsibles, compressing information into small icons – stars, knives, etc. – which are 'unfolded' to their full meaning by the informed reader: eg. 'one of the best tables in France, worth a special journey.'

Pop-up books leave behind the flatlands to which conventional books are confined. The *Bouncing Bugs* series by David Hawcock and Lee Montgomery, published by Random House, goes even further. These books turn into larger-than-life insects that can be suspended from the ceiling.

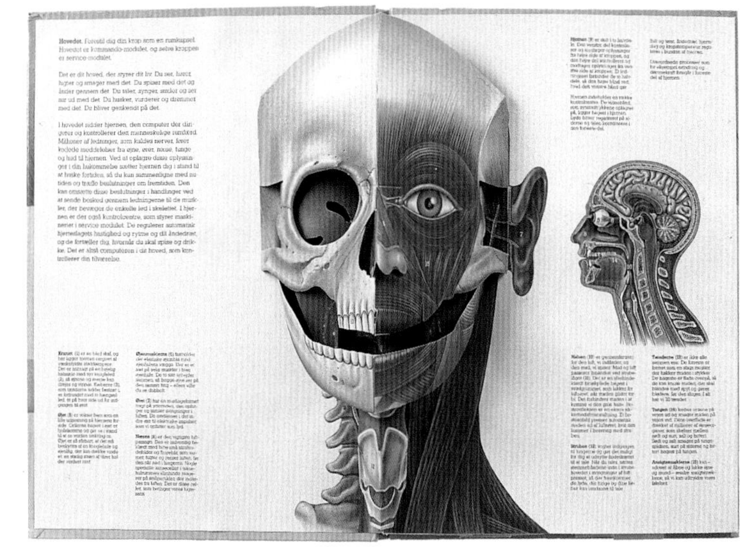

This Danish edition of *The Human Body,* published by Lindhart og Ringhof, projects from the page in unnervingly in-your-face, three-dimensional detail.

Newspapers come in different shapes and sizes, and these days also in many different colours, literally as well as politically. In format they cluster loosely around two generic types: the broadsheet and the tabloid. The tabloid is roughly half the size of its broadsheet big brother, and is generally considered less serious, especially when it comes with poster layout – ie with no body text on the front page. However, the tabloid has some practical advantages. It is much easier to handle, especially on crowded public transport. Editors, advertisers and production people alike all love its flexibility. And in any case, tabloid and broadsheet are equally worthless the next morning.

Whatever the format, newspapers are too large and floppy for convenient transport and storage without some sort of size reduction. They need folding. Producers and readers of newspapers apply up to three different principles of collapsibility. First, newspapers are produced as a number of *creased* sheets. Broadsheets are additionally cross-folded, leaving a permanent crease that halves the format. Second, newspapers may be *folded* once or twice more to fit the mailbox, pocket or briefcase. Third, newspapers may be *rolled* to be delivered in a wrapper. Or they may be rolled on a stick if the subscriber happens to be a café or a library reading room (see p. 126).

◀
Novelty decorations of many kinds are based on pleats designs. If a party is not too wild the pleated paper lanterns and moons can be folded flat for storage and future reuse.

▲
Oriental paper lightshades made from flimsy paper on a skeleton of fine wire are widely available in many designs at low prices. Buy them folded flat, expand them and hang.

◀
This rugby ball is one of thousands of different paper sculptures for party decoration and window dressing produced by Paper Fantasies of Denmark.

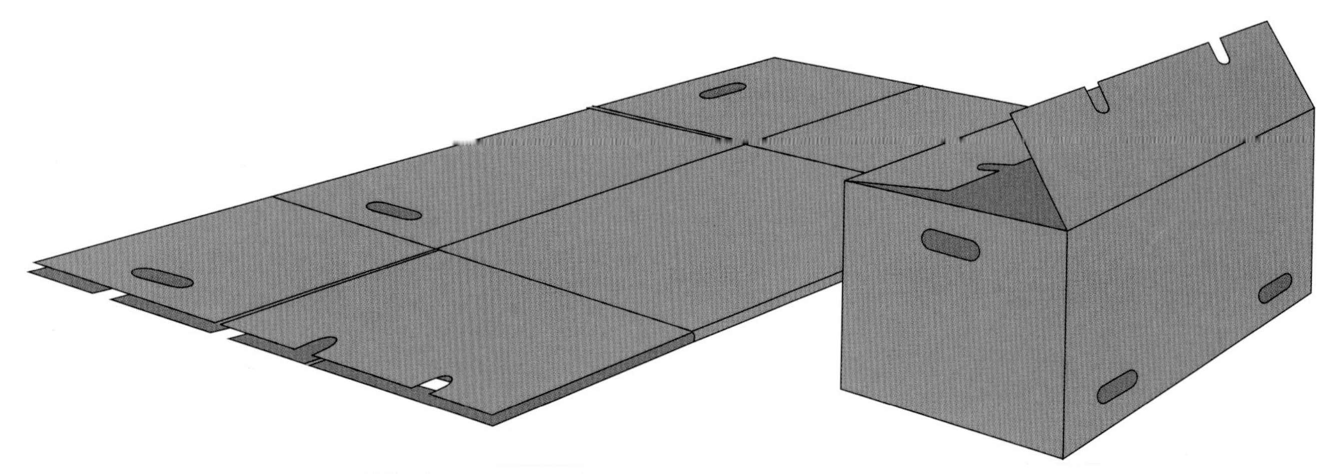

Utilitarian creases: generic cardboard box that folds and unfolds along prescored lines.

Say it with a see-through collapsible plastic vase.

Storage box sold by Muji in the UK. It is assembled with velcro and itself folds flat for storage.

Nomadic Seating of pleated cardboard designed by Teresa Sapey for the Absolut Vodka exhibition stand at the Milan Furniture Fair, 2000. It's a bold pleats concept, although hardly a hardcore again-and-again collapsible.

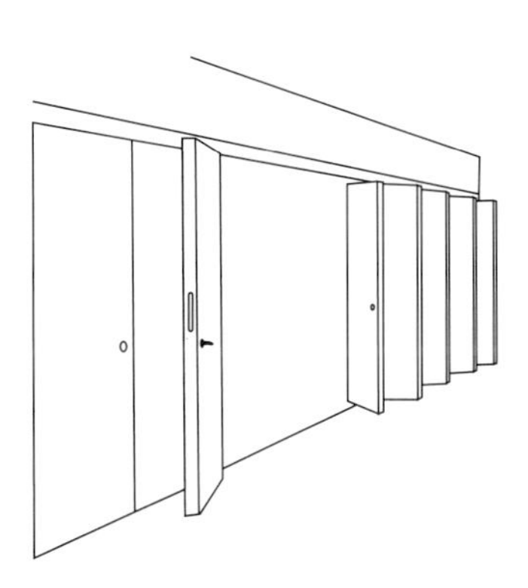

Foldaway walls often use the pleats principle, as here.

'Pleats please' sounded the mantra from Issey Miyake when in 1993 he introduced his latest marriage between traditional techniques and advanced technology. Ultra-light, soft and laundry-friendly, his synthetic materials were developed for polychrome orgies of pleats, pleats and more pleats. They are marketed through over a hundred Issey Miyake boutiques worldwide, some dedicated to pleats.

Bellows

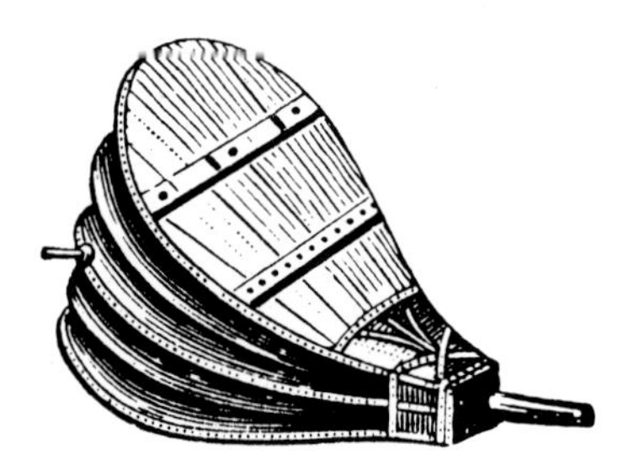

The traditional blacksmith's bellows is only a quasi-collapsible. Like a pair of scissors, its folded state is both active and passive.

A bee keeper's smoker, another quasi-collapsible, pumps smoke rather than air. This one was sold through the 1908 Sears, Roebuck & Co. catalogue, US.

A pair of bellows, the blacksmith's traditional instrument for aerating his forge, is essentially a contractible and expandable bag with sides folded in pleats.

As a mechanical device, a bellows is used where a flexible and sealed connection is needed between two planes. That flexibility may allow the object to be used as a pump, or it may simply make it adjustable in active use. A bellows may or may not be a deliberate means to reduce the object for storage. Thus although the basic principle is more or less the same across a wide range of applications, not all bellows are genuine collapsibles in our sense of the word.

The blacksmith's bellows and the bee keeper's smoke generator are air pumps. So are the accordion and concertina. These are all quasi-collapsibles: their collapsed condition is both a passive and an active state. Train wagons and articulated buses connected by corrugated bellows are adjustables rather than collapsibles.

On the other hand, cameras that use a bellows to create an adjustable sealed passage between lens and film, and also to reduce camera size when not in use, *are* genuine collapsibles. So too are airport bellows gates, which are contracted when not needed.

In the toolmaker's workshop illustrated in Diderot's *Encyclopédie* of 1763 the bellows is operated by pulling a chain. It is a quasi-collapsible.

Musical bellows: an accordion is a quasi-collapsible in which the collapsed state is both passive and active. The Gola accordion, made by German firm Hohner in the early 1950s, was named after the legendary accordion builder Master Giovanni Gola. Its bellows is made of manila cardboard lined inside and out with linen for 'optimal air management'.

An airport bellows gate, for weather-proof, vertigo-free access to the plane.

Collapsible modesty: a member of the NASA Skylab crew in a bellows shower cabin, 1973.

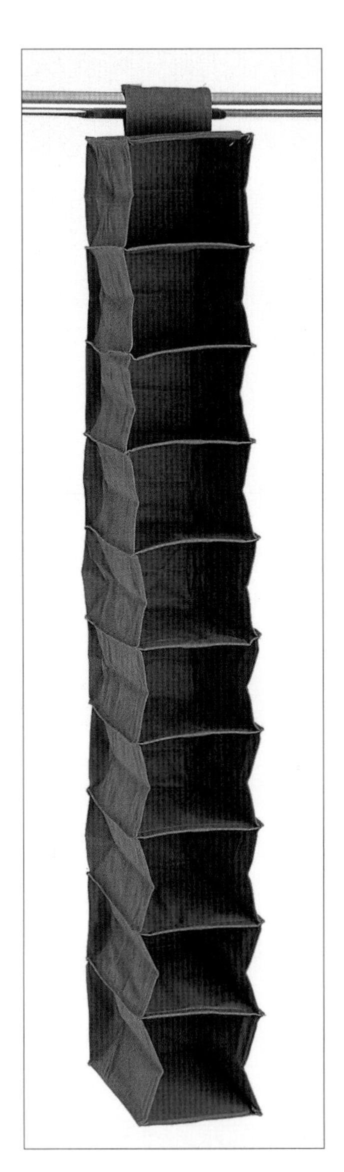

Shoe-shelving: this bellows collapsible rack hangs from the bar in a wardrobe to keep your footwear collection in order. Sold by Crate & Barrel, US.

Japanese girls with a traditional collapsible lantern. *The Hour of the Cock* from *Twelve Hours of the Green Houses*, Utamaro, *c*. 1795.

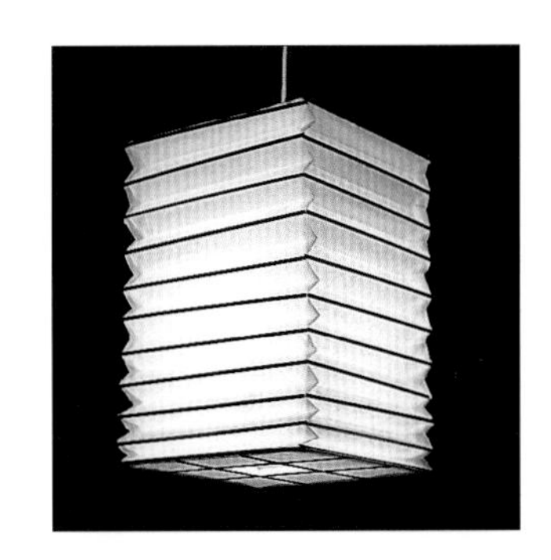

Oriental paper lamps are a cheap, cheerful and widely available application of the bellows principle.

VEST POCKET HAWK-EYE FOLDING CAMERA

Takes Pictures 1⅝ x 2½ Inches

Made by the Eastman Kodak Company

Hardly a Handful, yet Takes Excellent Pictures

This camera is "hardly a handful"—yet it makes sparkling, clear-cut pictures, in the ever popular "vest pocket" size—1⅝ x 2½ inches. It actually fits into any pocket, into the vacation grip or the handbag, yet it is as dependable a picture-maker as you could want—always ready in a wink for the chance snapshot—always sure to get a good one if you will give it half a chance. The pictures are good, so good in fact, that you can have them enlarged.

All this means, of course, that the Vest Pocket Hawk-Eye is both simply and scientifically constructed. The focus is "fixed"—draw out the bellows and you are ready to make the picture. The shutter gives a choice of either instantaneous or time exposures. The brilliant finder, reversible for horizontal or vertical views, makes it easy to sight the subject. A tripod socket and easy daylight loading are other conveniences. Uses Ansco, Eastman or other standard roll film, 8 exposures. Every purchaser of a Hawk-Eye Camera is entitled to a twelve months' subscription to "*Kodakery*." Simply fill out the subscription blank you will receive with the Camera, mail it in direct to the Eastman Company, and you will receive the magazine free for one year. It is interesting and instructive to young and old alike. It will double the fun you'll have with your Camera.

No. 4174. Vest Pocket Hawk-Eye Folding Camera, Picture Size 1⅝ x 2½ in. Price..........$5.00
No. 4175. Films for No. 4174 Camera (8 Exposures). Price Postpaid..........25

FOLDING CARTRIDGE HAWK-EYE CAMERA

THREE POPULAR SIZES
2¼ x 3¼, 2½ x 4¼ and 3¼ x 5½

All made by the famous Eastman Kodak Co.

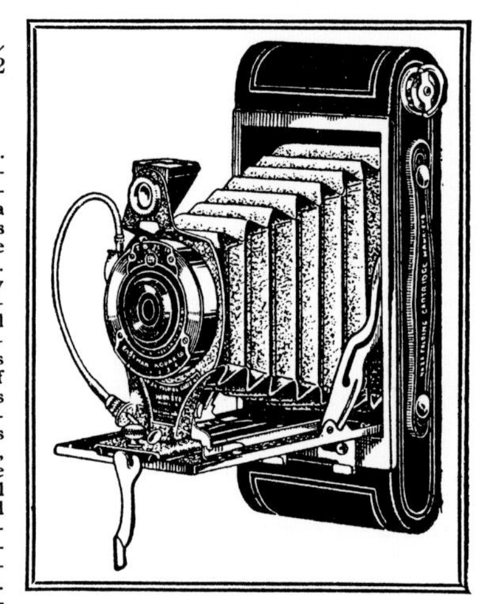

These three Cameras are all alike except in size. All three are small enough to fit in the pocket, inexpensive enough for almost any purse—yet complete and proficient in every respect. There is a handy scale which gives adjustments for various distances. But the lens can also be set so that the camera may be used as a "*fixed-focus*" outfit. The manual of directions tells how. But this is only one instance of the camera's versatility. The brilliant View Finder can be set for either horizontal or vertical views. The Kodex Shutter, with snap-shot speeds of 1/25 and 1/50 second, besides bulb and time, makes possible a variety of exposures to fit various light conditions. Exposures are made either with a lever or with the cable release supplied with each camera. The cameras are covered with fine grain black imitation leather, handsomely embossed. Exposed metal parts are trimmed in durable black crystal lacquer, nickel and brass. In short the Hawk-Eyes are built and equipped to make good pictures, and they are finished to present an appearance that is to be expected in a much higher priced Camera. All Hawk-Eye Cameras are Eastman made from lens to lock. That fact insures quality material and workmanship—and what is still more important, good pictures. Each Hawk-Eye camera carries with it a year's free subscription to *Kodakery*, an instructive, well written, finely printed and profusely illustrated photographic monthly that sells regularly at 60c per year. You merely fill in the subscription blank on the first page of the instruction book that accompanies your camera, mail it in to the Eastman Kodak Company (not to us) and you will receive the magazine absolutely free, as stated, for twelve months.

No. 4176. Folding Cartridge Hawkeye Camera. Size of Picture 2¼ x 3¼ Price..........$8.00
No. 4177. " " " " " " " 2½ x 4¼ Price..........9.00
No. 4178. " " " " " " " 3¼ x 5½ (Post Card size) 13.50
No. 4171. Films for No. 4176 Camera, (6 Exposures). Price Postpaid..........25
No. 4173. " " " 4177 " (6 "). Price Postpaid..........30
No. 4179 " " " 4178 " (6 "). Price Postpaid..........55

JOHNSON SMITH & CO., RACINE, WIS. **163**

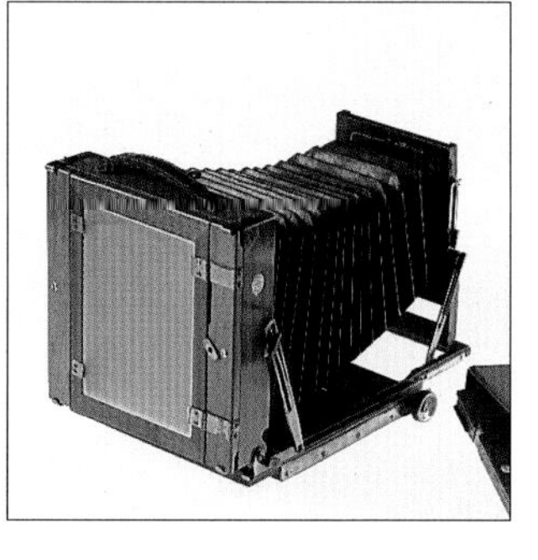

Many cameras have a sealed, light-proof bellows to create an adjustable passage between the lens and film, and to allow the camera to be folded, if not quite to pocket size, then to a more convenient size and format for carrying. The triple-extension Imperial camera launched by Thornton Pickard of the UK in 1903 (above) and the Hawk-Eye folding cameras made by the Eastman Kodak Co. of the US in the 1920s (left), all had bellows.

Weighing in at 2 kilos (4 lb 6 oz), the Polaroid Land camera introduced the 60-second all-in-one photographic process to the market in 1948. Edwin H. Land invented the process, and Walter Dorwin Teague created the final design. The Polaroid Corporation was named after another Edwin H. Land first: light-polarizing filters.

Markies (Dutch for 'awning') is a caravan that folds out by a motorized device to triple its floor area. The side walls descend to create two new zones for living. One zone, the living room/terrace, is covered by a transparent awning which in fine weather can be left open. The other zone, the bedroom, is covered by an opaque awning, and can be split into smaller units. Kitchen, dining space and wc are in the solid central area.

The caravan was designed by Eduard Böhtlink, 1986–95.

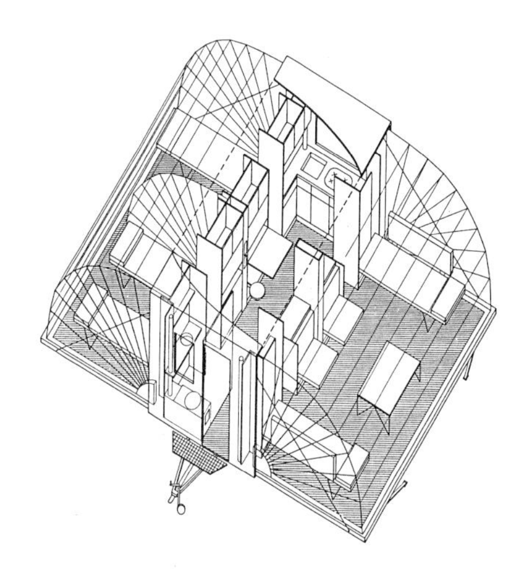

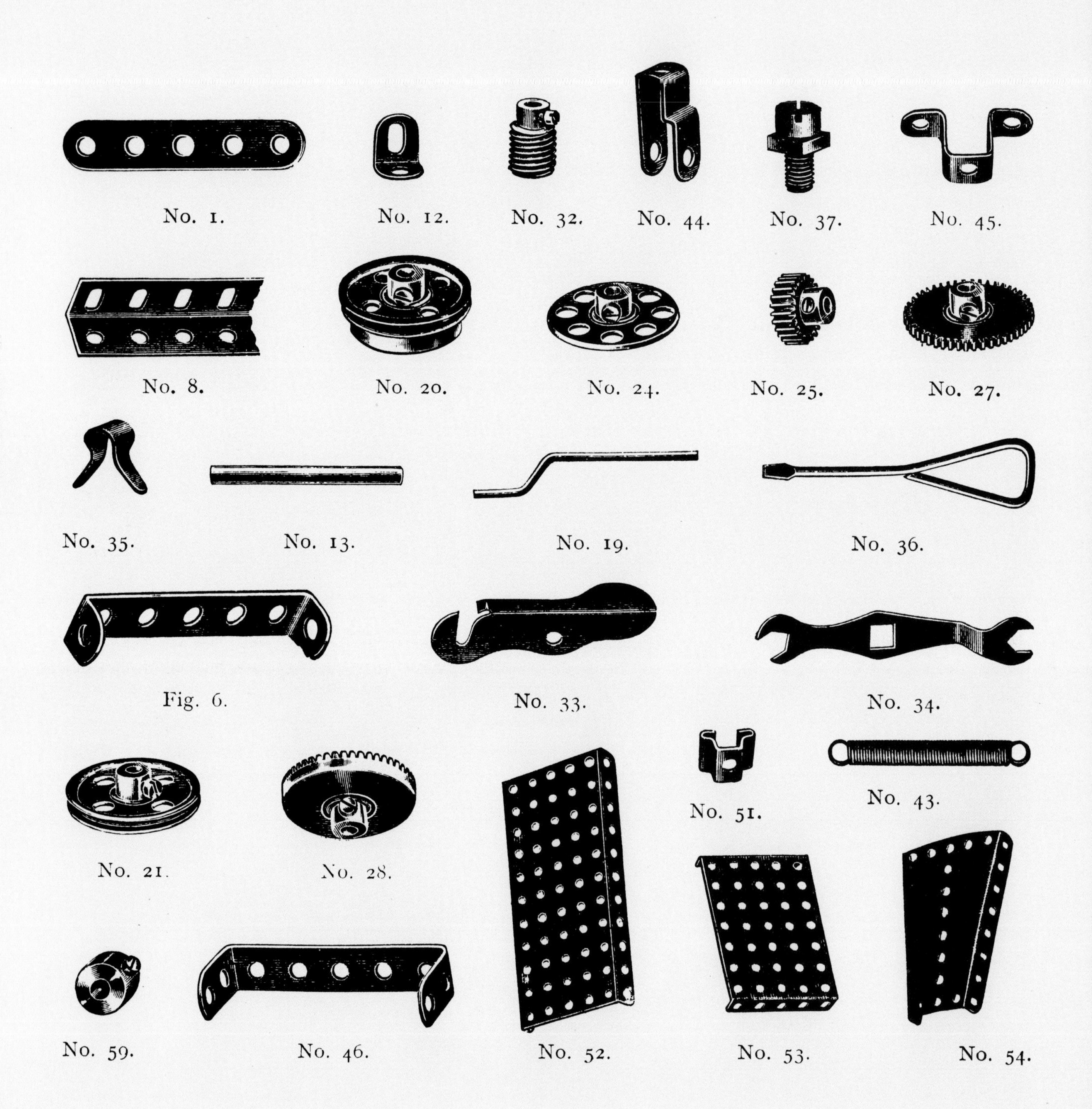
No. 1.
No. 12.
No. 32.
No. 44.
No. 37.
No. 45.
No. 8.
No. 20.
No. 24.
No. 25.
No. 27.
No. 35.
No. 13.
No. 19.
No. 36.
Fig. 6.
No. 33.
No. 34.
No. 51.
No. 43.
No. 21.
No. 28.
No. 59.
No. 46.
No. 52.
No. 53.
No. 54.

Assembling

◄
The Meccano construction kit was always more than the sum of its parts. In its heyday, it was not just an educational toy but a worldwide movement.

To assemble a number of separate parts into a whole, and then later to dismantle that whole again into its parts for storage, is to practise a time-honoured principle of collapsibility.

Children's construction kits and toy bricks such as Meccano and Lego are an early encounter with the assembling principle that some people never quite grow out of. But adults do not only sneakily play with their kids' toys; they have their own construction kits too. The FAC system was designed for grown-ups to build function models. Other adult construction kits include temporary scaffolding and a number of exhibition systems.

Trainee soldiers learn to assemble and dismantle mechanical implements from weapons and tents to bridges and vehicles, if necessary in total darkness. Marksmen, snipers and undercover agents assemble their lethal precision weapons before taking aim.

Novice campers often liken their newly acquired tent to a puzzle as they struggle to identify the nature and purpose of its various parts. How do they connect? Jigsaw puzzles are two-dimensional examples of the assembling principle. As with other pastimes, the journey not the destination is the purpose – process over product.

Two thin wooden plates with pre-cut push-out parts assemble into an authentically scary spinosaurus skeleton.

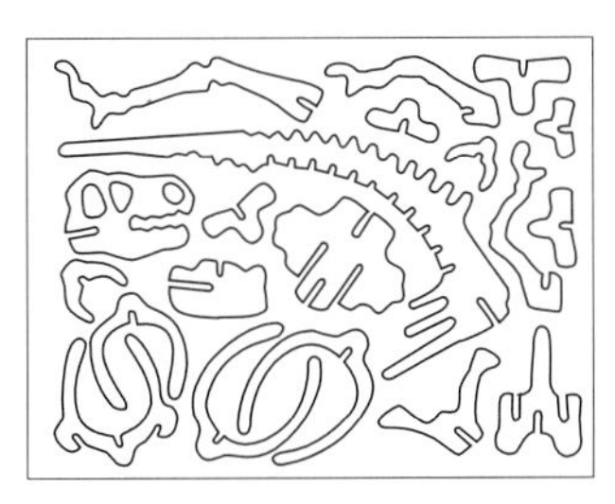

Every adult remembers their favourite toy, and it will often be a set of wooden bricks that combine in an unlimited number of different ways. The Swiss toy manufacturer Naef specializes in wooden toys that challenge children's skills of combination – while pleasing the aesthetic sense of their parents. Cubicus designed by Peer Clahsen in 1967, and the Naef Game by Kurt Naef in 1962 (above left and right), are colourful cases in point.

Jigsaw puzzles are absorbing two-dimensional collapsibles. As a jigsaw, Juan Miró's surrealist painting *Person with Eye and Foot* (1938) puzzles the user after as much as during assembly. It is sold by the Miró Museum, Barcelona.

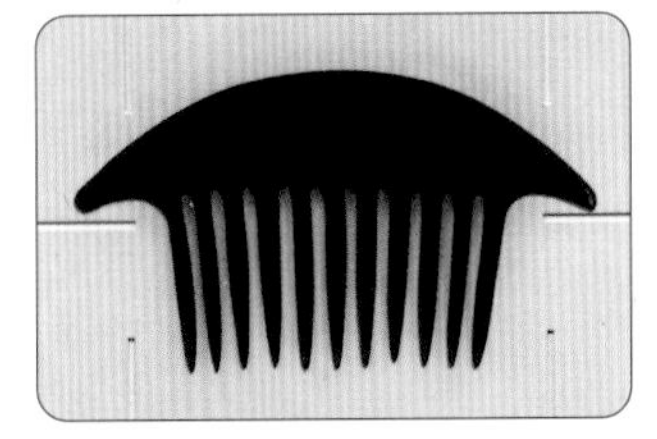

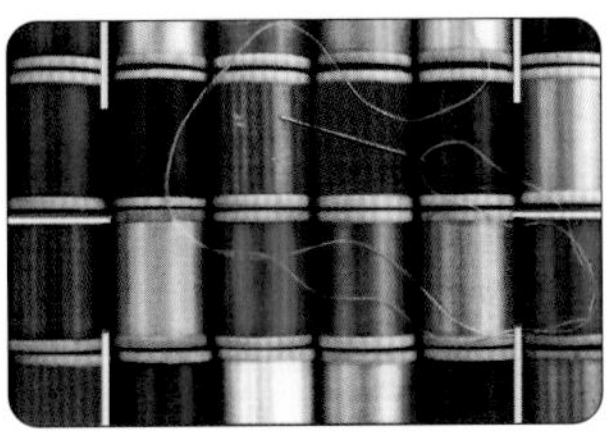

Charles Eames reportedly thought of the toys he designed as gifts for his grandchildren and the children of his friends. His House of Cards of 1952 is yet another entry in that category of toys that are appreciated as much by parents as by their offspring. The original House of Cards comprised two decks, one illustrating textures, the other 'good things'. The deck produced by Otto Maier Verlag of Germany today is a combination of these original two. In 1953 Eames designed the Giant House of Cards, twenty 18 x 28 cm (7 x 11 in) plywood cards, and in 1970 a Computer House of Cards for IBM.

Meccano is probably the best-known recreational construction system ever. In the 1930s, boys in many different countries built Meccano models, kept their Meccano promises, joined the Meccano Guild and received the Meccano Magazine eight times a year. The kit was designed in 1901 by Frank Hornby of Liverpool, who called it at first 'Mechanics Made Easy'. Specialist sets were produced over the years, but the basic kit always consisted of perforated metal strips and plates, plus nuts and bolts with which to fasten them together. The idea was to prepare boys for working as engineers by enabling them to build models that worked like the real thing. Many years later, when the popularity of Meccano waned, its share of the toy market was usurped by models that looked more like the real thing; appearance over mechanics.

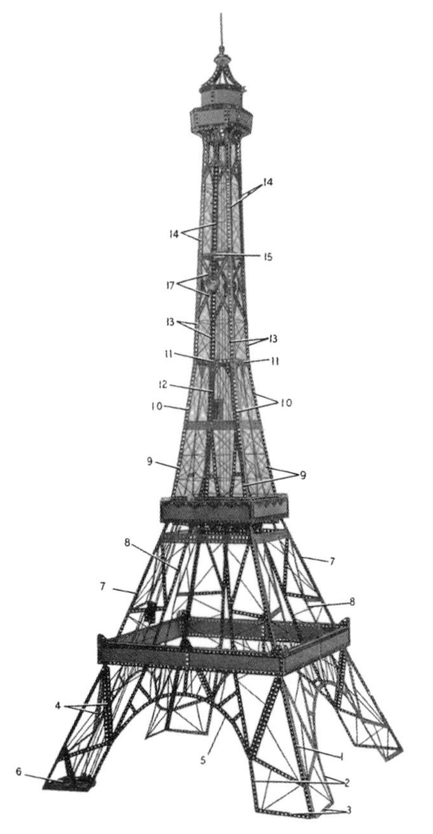

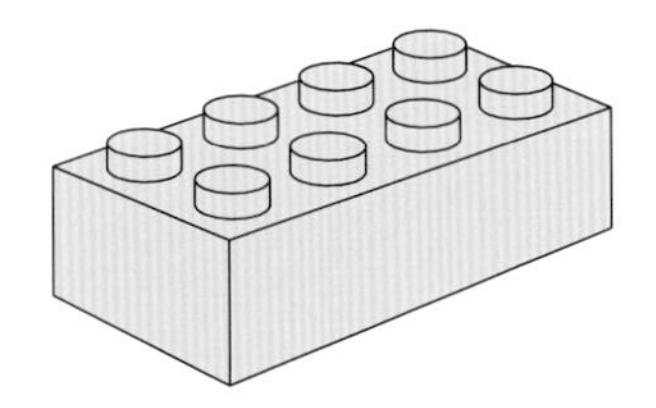

The word Lego was coined as an abbreviation of '*leg godt*', Danish for 'play well'. Only later did the company happily realize that 'lego' in Latin means 'I combine'. The basic Lego building block is a plastic brick with eight knobs on one face. These units combine in a vast multitude of ways – exactly how many could be calculated by a mathematician. Lego bricks are sold in basic sets from which builders can construct any number of objects, and dedicated sets to build specific objects. Mindstorm, the much talked-about marriage between Lego bricks and computerized electronics is the latest development. 'If you can't beat the PCs, deal with them' the toy brick people argue.

Many male travellers have experienced forgetting their shaving gear. At the Hotel New York in Rotterdam, irritation turns to pleasant surprise when they are offered this highly collapsible personal tool kit.

The Evantgarde motorized wheel-chair collapses partly by folding, partly by dismantling, for carrying in the boot of your car. Manufactured by Otto Bock, Germany

The FAC system, designed by the Swede Mark Sylwan in 1952, was originally aimed at model builders of more advanced skills who had outgrown their Meccano sets. Only later was FAC's potential on the professional market realized. Today, FAC is sold as loose parts as well as in four standard sets: Training Equipment, Standard Equipment, Gear Equipment and Structural Equipment.

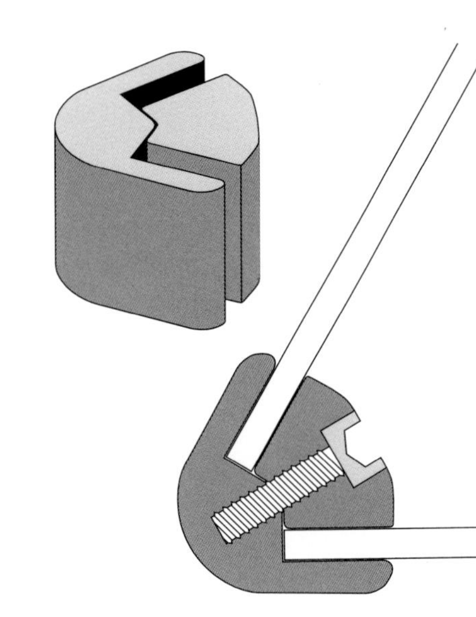

The Mero Voluma construction system connects panels for displays and partitions at temporary exhibition sites. An allen key is the only tool needed for assembling and dismantling the whole show. If God is in the details, as Mies van der Rohe believed, then He would be just about here. Manufactured by MERO-Raumstruktur, Germany

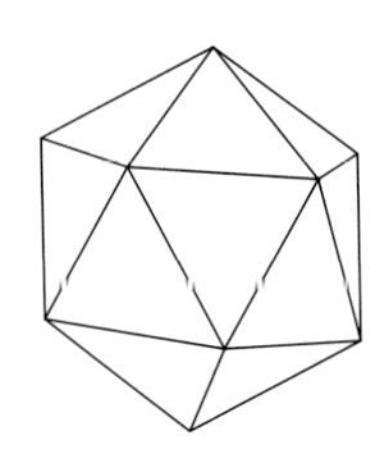

An icosahedron is a regular twenty-sided polyhedron. If its faces are subdivided further, the polyhedron will gradually approach the form of a sphere. Meroform domes for exhibitions and other temporary purposes are hemispheres based on subdivided icosahedra built of nodes and tubes. Meroform standard domes are available with diameters from 5 to 16 metres (16 to 52 feet). An extra ring of faces at the base adds to the overall height of the domes.

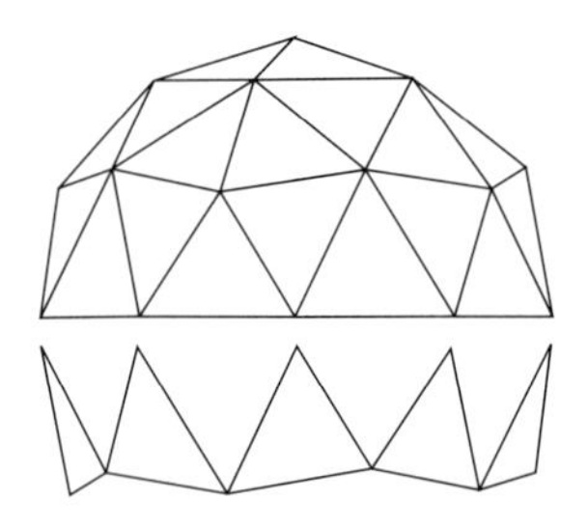

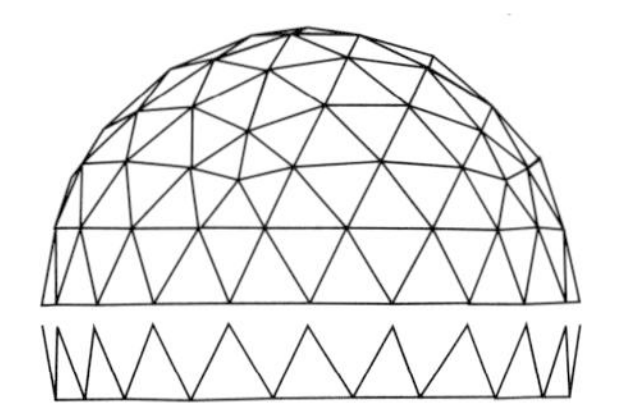

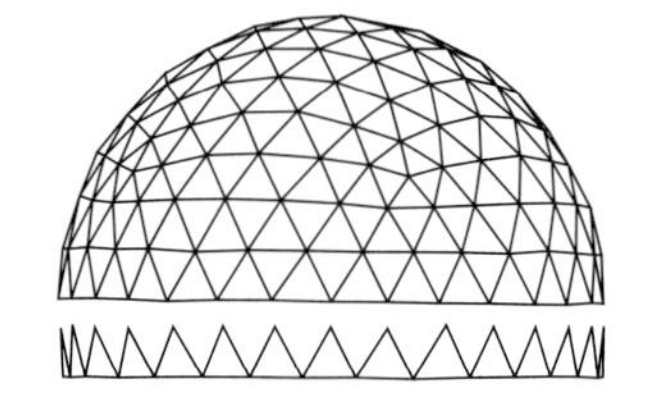

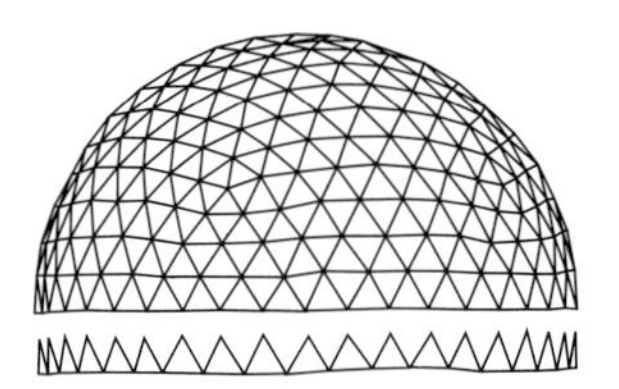

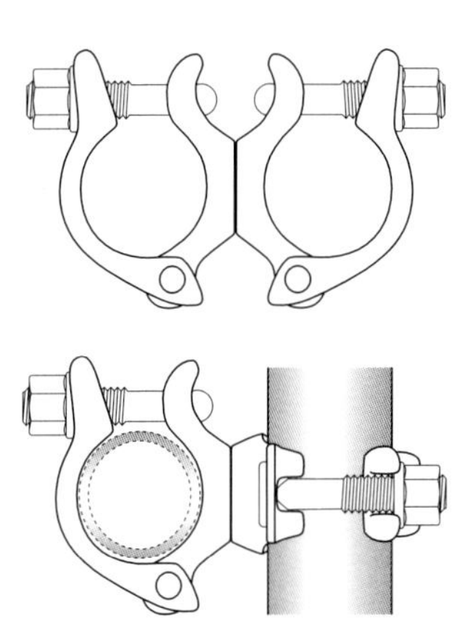

Scaffolding is assembled from rods and joints, most commonly key clamps such as these.

Built in 1888 for an industrial exhibition in Copenhagen, this 26-metre (85-foot) beer bottle stands today at Strandvejen in Hellerup, north of Copenhagen. The temporary scaffolding was erected for repair work. And in answer to the number one FAQ – it would hold the equivalent of one and a half million regular-sized bottles of Tuborg beer.

Cranes are dismantled into truck-sized parts to be moved on to the next building site and reassembled. For builders, mastering the logistics of the heavy equipment is a survival discipline.

'Bailey bridging made an immense contribution towards final victory in World War II. As far as my own operations were concerned with the Eighth Army in Italy and with 21 Army Group in northwest Europe, I could never have maintained the speed and tempo of forward movement without large supplies of Bailey bridging.'
Field Marshal Montgomery

The Bailey Bridge is named after its British inventor, Sir Donald C. Bailey. It is a prefabricated bridge constructed from a small number of standard components. The basic element is a lattice-work panel. A number of these are fastened together end to end by bolts and are locked by split pins. Two lines of panels extending the full length of the bridge are connected by steel beams which carry the roadway of wood or other material. The bridge is constructed on one bank and then moved forward on rollers over the gap until it meets the other side. The whole thing can be assembled by a crew of forty trained sappers armed with nothing more than spanners. One panel can be carried by six sappers and all parts fit into a standard three-ton lorry. A Bailey Bridge will carry heavy loads, and can be further strengthened by adding more steel girders.

Hinging

Hinging is the *primus inter pares* of collapsibility principles. The term hinge covers a wide spectrum of flexible joints, from the long hinge joining lid and body of a Steinway grand piano to the multiple small joints connecting a folding ruler.

While a hinge traditionally consists of two or more moving metal parts, modern hinges are often made of a single piece of plastic that bends repeatedly. Thus the boundary between creasing and hinging – like that between creasing and folding – is sometimes blurred.

The supporting structure of umbrellas is hinged. So too are the legs of a collapsible tripod, though they may additionally expand and collapse by sliding. Hinging has long been the most commonly applied collapsibility principle in furniture (see p. 174). The screen and the keyboard of a laptop computer are connected by hinges, as are legions of man's other more or less indispensable tools. The importance of being hinged is eloquently captured in the phrase 'he is unhinged'.

The author's best friend: an Apple Macintosh PowerBook with hinged screen and keyboard. Laptop computer technology is steadily approaching its ultimate ideal: an ultra-flat, totally collapsible keyboard (see pp. 91 and 131), together with a free-floating screen image, releasing us from the hardware orgies of the past.

The Steinway & Sons grand, concert-hall essential and living-room jewel. This harmonious hinged collapsible weighs half a ton and its parts number 12,000.

Hinging as a unique selling point: the TS 505 portable radio designed by Marco Zanuso and Richard Sapper in 1964. Manufactured by Brionvega, Italy.

More than a mobile phone: the R380 Smartphone from Ericsson, Sweden.

More than a mobile phone: the Communicator from Nokia, Finland.

The hinged satellite antenna of the Capsat mobile phone works in the most remote parts of the world via the Inmarset Satellite Network. The phone weighs 2.2 kilos (4 lb 13 oz) and is produced by Danish firm Thrane & Thrane.

Relentless pressure to miniaturize can go too far. Screens become too small for the human eye, buttons and knobs too small for adult hands. While many handheld computers are just about big enough to read, they are extremely difficult to write on. The work area is just too small. The foldable Stowaway keyboard devised by Think Outside in the US solves that problem. It folds down to the same size as the palm computer and folds out to a full-size keyboard for full-size fingers.

The Polaroid SX-70 from 1972 included a number of improvements on the original 1948 model (see p. 73). Among these is the space-saving SLR (single lens reflex) and the power source included in the film pack. Dr H. Land was the inventor, with Henry Dreyfuss Associates acting as design consultants.

The item most often left behind on the New York subway derives its name from the Latin *umbraculum*, meaning little shadow. Originally – in Egypt, Assyria, Persia, China and elsewhere – the umbrella protected its owner against the sun. It was also a symbol of high office. In 1750 Josiah Hanway introduced the umbrella to the British as a device for protection against rain, and they took it up with great enthusiasm. Few remember Hanway, though his name is immortalized in the name of a London street.

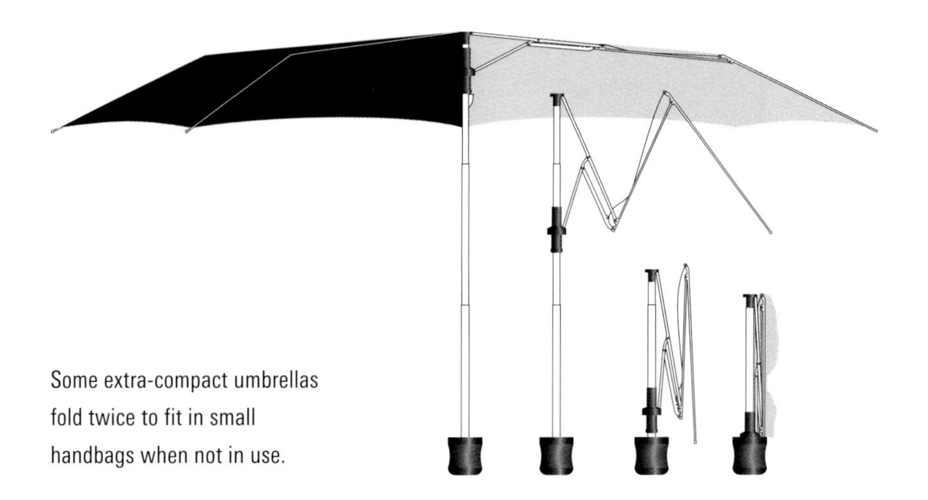

Some extra-compact umbrellas fold twice to fit in small handbags when not in use.

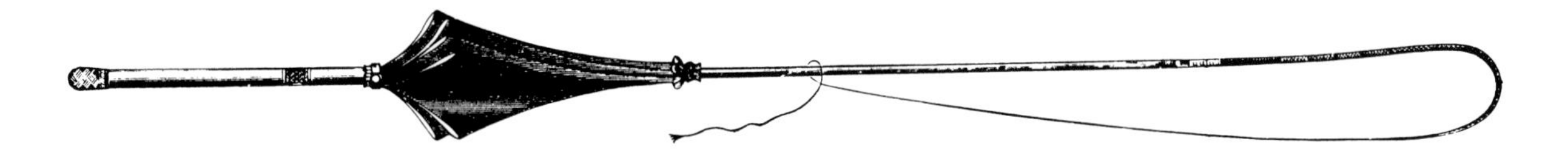

Lady's parasol-whip, offering sun protection and horse motivation in one tool. It was sold in the US in the 1890s through the C. M. Moseman & Brother catalogue.

An umbrella that truly deserves the name umbrella is called a parasol.

Utility with elegance: a mid-19th century Shaker swift, an expandable hinged circular frame used for winding skeins of yarn.

Utility without much claim to elegance: a collapsible clothes-drying rack formed of rods that extend on flexible joints.

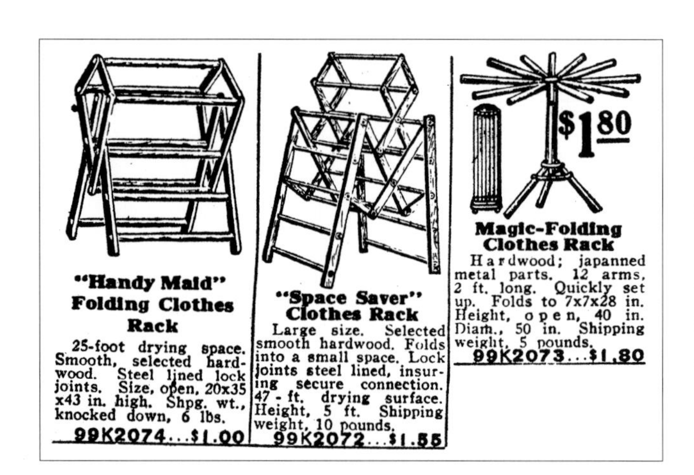

The Handy Maid, Space Saver and Magic Folding clothes racks, advertised in the American Sears, Robuck & Co. catalogue in 1927.

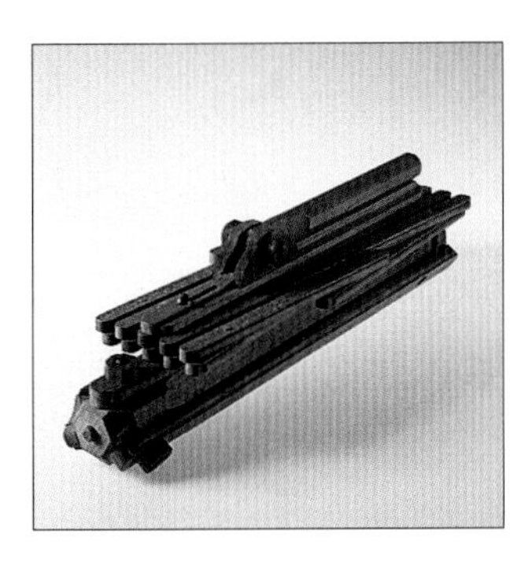

Collapsible sheet-music stand suitable for woodwinds and strings. It is height-adjustable from 44 to 132 cm (17 to 52 in). Apart from hinging, it also applies the principles of assembling and sliding. Manufactured by Andreas Mannhardt, Germany.

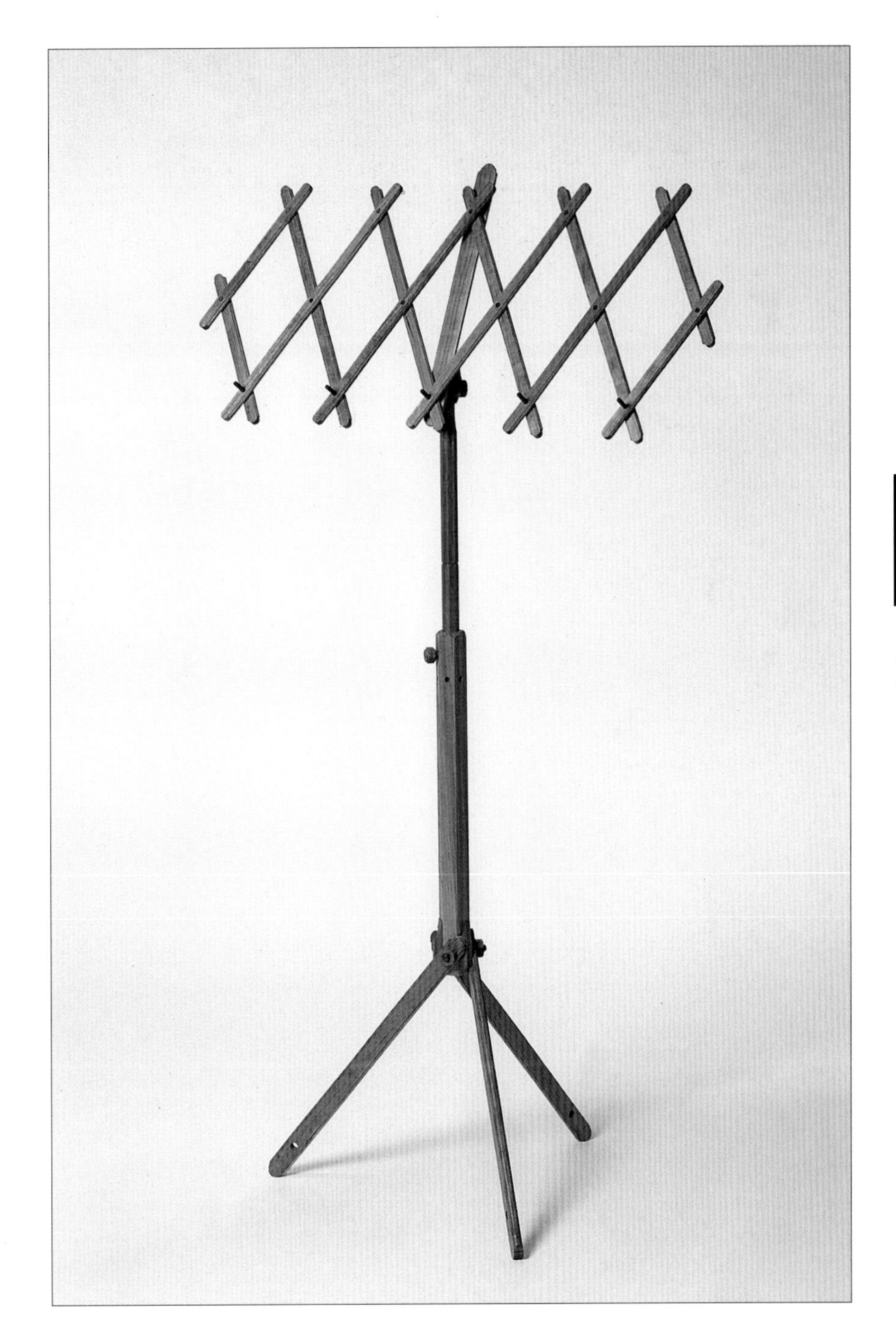

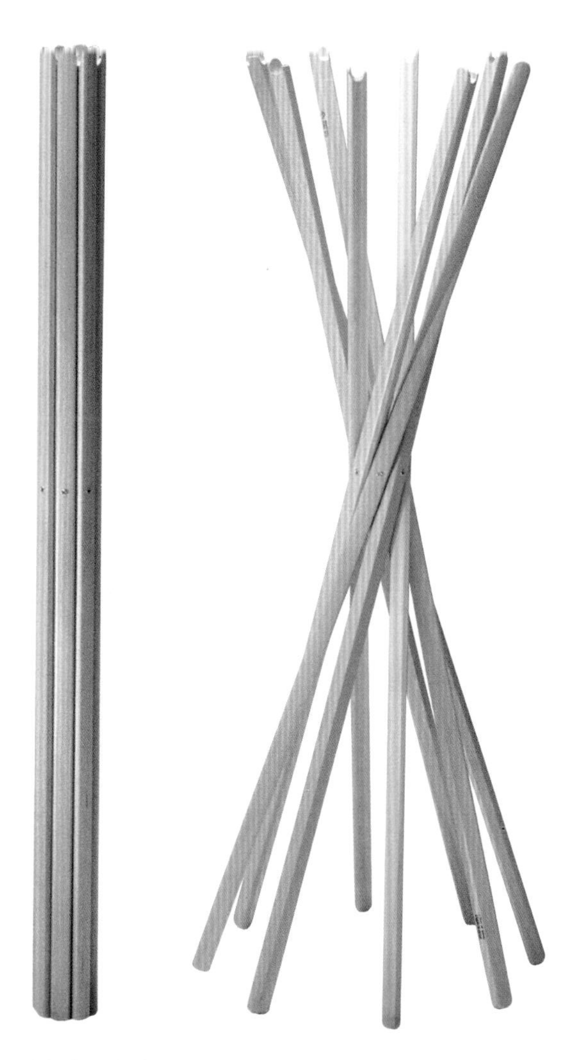

Sciangai, Italian for 'Shanghai', a hat and clothes stand designed by the triad of De Pas, D'Urbino and Lomazzi in 1973. Manufactured by Zanotta, Italy.

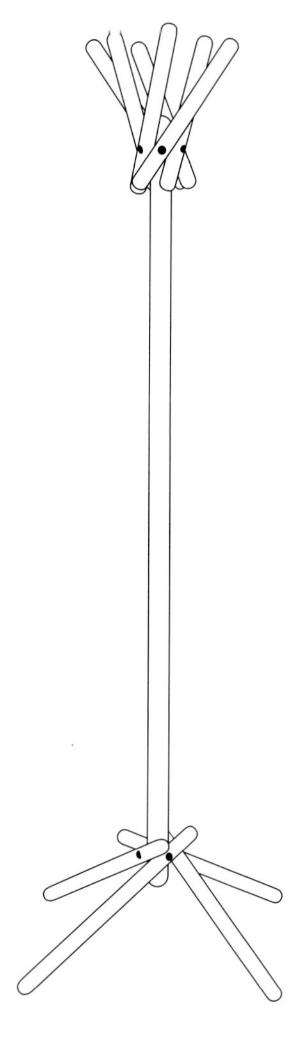

Broomstick clothes stand designed by Vico Magistretti in 1978–79. Manufactured by Alias, Italy.

STUDIO EASELS

(A) (C) (B) (D)

(A) Hardwood Studio Easel
This easel's for you . . . if you want a durable, beautiful tool that will provide a lifetime of dependable service. 86" high x 26" wide, it is carefully constructed of clear-lacquered, fine-grained hardwood. The 26" wide box-type utility tray has both a palette holder and a 35" canvas support and is movable on the one-piece vertical steel ratchet. A movable canvas holder allows you to mount up to a 60" canvas. Four casters (two are locking.)

Number	Each	2/Ea	6/Ea
983100	$174.00	$159.00	$143.00

(B) Hardwood Studio Easel
This ever-popular, space-saving natural hardwood easel has a clear lacquered finish. The generous 48" center post ratchet will lock the combination canvas tray and palette holder in any position you need. Base stabilizers, no-skid rubber bumpers, plated hardware and a flat-folding rear leg complete this easel's features. 78" high x 28" wide.

Number	Each	3/Ea	6/Ea
980000	$81.60	$75.40	$67.80

(C) The Greenwich from Anco
The top-of-the-line professional studio easel made of select lumber with natural lacquer finish. The Greenwich features an extra oversize reversible canvas tray which functions independently of the screw-wind adjustment. It may be inverted for use as a canvas holder. The Greenwich is mounted on four locking casters and may be tilted forward or back. Material tray is 26¾" wide x 4¾" high x 4⅜" deep and can be raised or lowered by smooth action hand crank. Takes up to 72" canvases. Minimum height: 8-ft, maximum height: 11½-ft. Base: 30¼" wide x 36" deep. Top ledge: 2½" between post and safety ledge. Shpg Wt 75 lbs. Truck shipment. See page 1.

Number	Each	2/Ea
F-983000	$339.90	$324.90

(D) Hardwood Studio Easel
A quality easel at a reasonable price! This fine-grained hardwood easel features a 36" movable ratchet that will lock the material tray in any position you need. The side supports add pleasing stability and the canvas holder, rubber bumpers, and a flat-folding adjustable rear leg are features that will make this lacquer-finished easel an economical but functional addition to your studio or classroom. 72" high x 28" wide.

Number	Each	3/Ea
983300	$94.60	$84.90

Our one toll-free number serves the entire country. Your order will be shipped from the location nearest you. 1-800-447-8192 In Illinois: 1-800-322-8183 For orders only, please.

94

STUDIO EASELS

(A) (B) (C) (D)

(A) Stanrite Aluminum Easel
No tools are needed to set up this extremely sturdy easel! Holds canvases up to 48" high. "Auto-lock" holds top of canvas securely. 21" wide tray holds canvas, painting knives, palette cups. Palette may be held by arms under tray. Rubber-tipped legs. Folds flat. Ships UPS.

Number	Each	3/Ea	6/Ea
848000	$61.90	$56.90	$50.80

(B) Anco Celebration Easel
The ultimate and most versatile studio easel ever. There is nothing like it anywhere. It will hold up to an 86" canvas; the adjustable screw-winding mechanism raises and lowers the canvas; and the frame-in-frame canvas rest tilts backwards and forwards, locking into multiple positions that are impossible on any other easel. Locking casters permit easy mobility. Multiple canvas holders provide for the utmost in flexibility. All hardware is brass-plated. Constructed of clear hardwood and the 4-coat finishing process gives it a superb look. So superb, people decorate and display paintings with it. Minimum height: 68". Maximum height: 105". Base: 30" x 35". Materials tray: 28" wide. Weight: 80 lbs. Truck shipment. See page 1.

Number	Sugg. List	Each
F-979300	$695.00	$490.00

(C) The Laguna
This Brazilian mahogany easel is the nicest portable model we've seen. Weighing just 7 lbs. and able to fold nearly flat for storage, it nevertheless has the rigidity and three-point stability to support good-sized canvases (up to 46"). Total height: 69". Ships UPS.

Number	Each
6886800	$37.80

(D) The Brasilia
A solid Brazilian mahogany easel that features twin-track design to hold up to a 60" canvas. The design of the accessory tray is superior to other easels because of its separation from the canvas holder and its hinged-door storage compartment. The Brasilia moves easily on two rear casters, but cupped and adjustable feet in the front assure stability. All in all, it is a nicely detailed easel at a price too good to pass up! Imported exclusively by Dick Blick. Total height 86". Ships UPS.

Number	Each
6886700	$198.00

95

A quality easel is no guarantee of quality art, though it can be a pleasure in its own right. These easels were demonstrated in the 1986 Dick Blick catalogue, US.

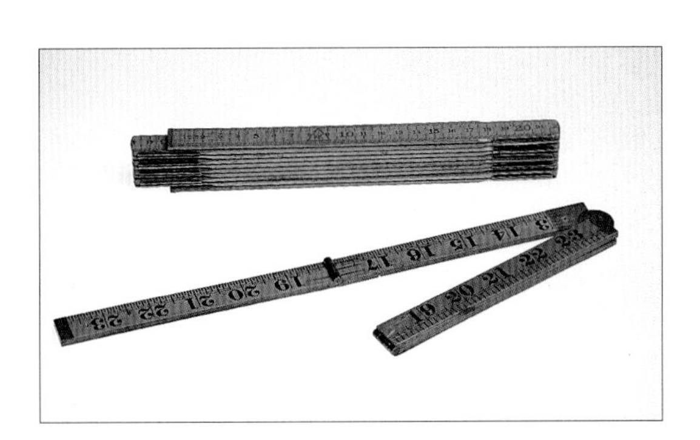

Collapsible rulers.

Step-ladders and mobile scaffolding are indispensable when you need them, yet cumbersome when not in use – making them obvious candidates for collapsibility.

Double-sided step-ladder made by Günzberger Steigtechnik, Germany.

Hinged step-ladder pictured in an 1881 housewares catalogue, US.

Mobile scaffold, also by Günzberger Steigtechnik.

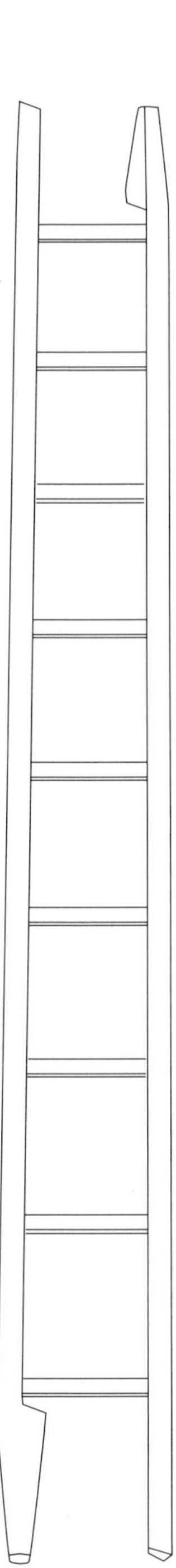

Fire-fighters' ladder that doubles in its folded state as a battering ram. Manufactured by Günzberger Steigtechnik.

Linen testers or folding magnifiers come in different materials, but the hinged design is generic.

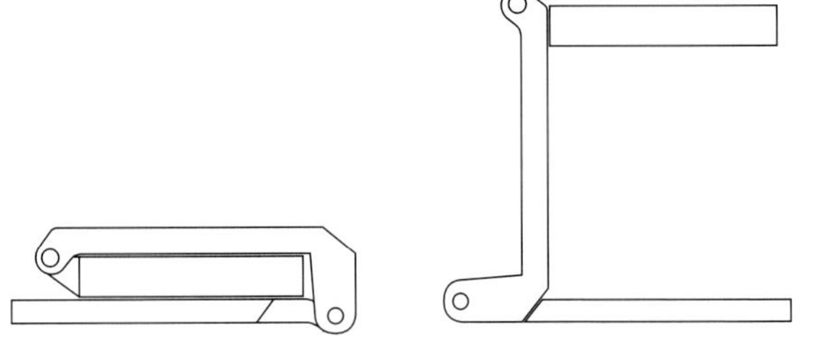

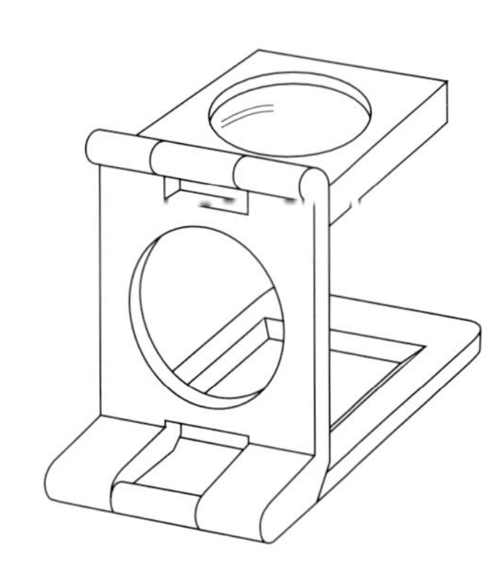

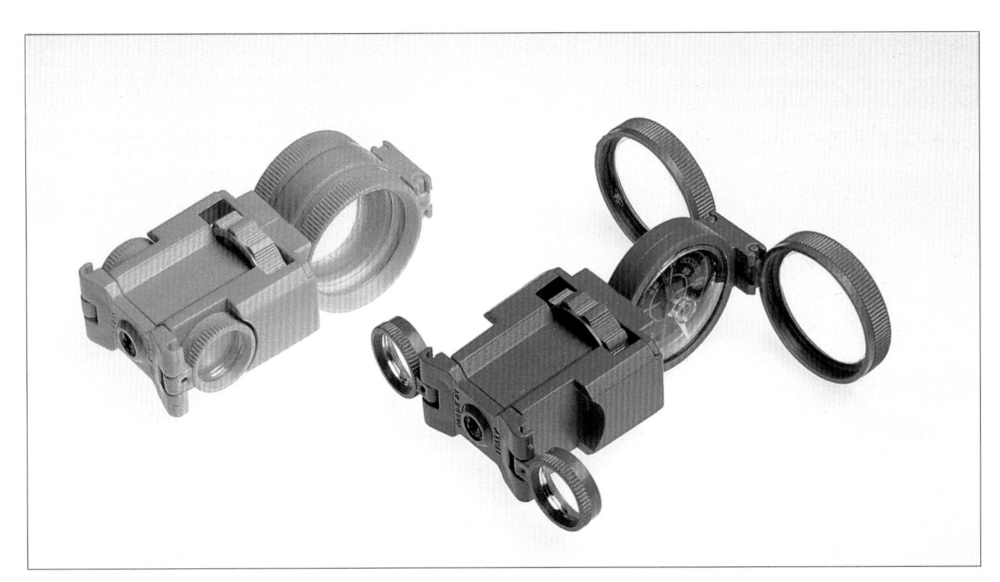

Optic Wonder is a multi-functional optical package: telescope, binoculars, magnifying glass, compass, signal mirror and stereoscopic viewer, plus some gadgets besides. But the concept is not entirely new. Manufactured by Navir, Italy.

The 9-in-1 Opera Glass Novelty advertised in the 1929 Johnson Smith & Co. catalogue, US.

Old British Army compass.

Compasses and compasses are two different things. A magnetic compass is a device for finding north, while a pair of compasses is a tool for drawing circles or transferring measurements. The former *may* be, the latter *always* is, a collapsible.

Compasses by Silva, Sweden.

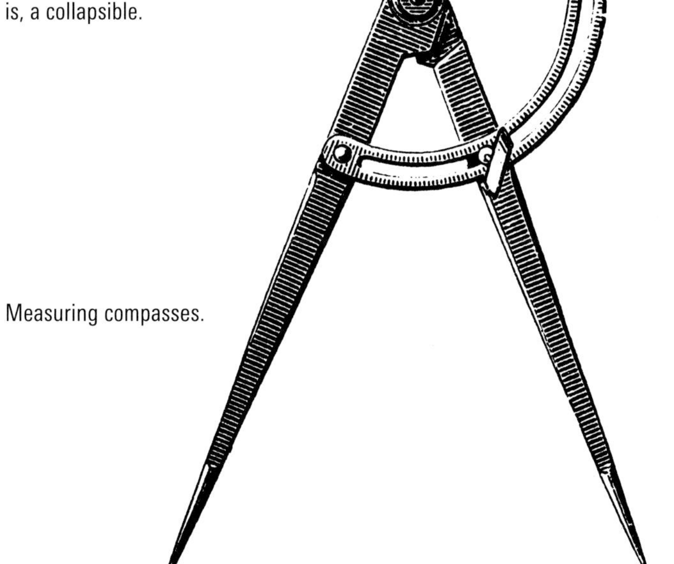

Measuring compasses.

'The better you look, the more you see,' says writer Bret Easton Ellis. As marketers and wearers of spectacles have become increasingly fashion conscious, glasses have become 'eyewear' and 'designer frames'. The optical technicalities are now as nothing to the fashion part of the business: what others see is more important than what you see yourself. If handicapped by *not* having impaired vision, you can always resort to sunglasses. Fashion or not, the basic design of spectacles is to a large degree determined by the nature and position of the eyes and the carrying capacity of ears and nose. Hence the fundamentals of spectacle design haven't changed much in the last hundred years.

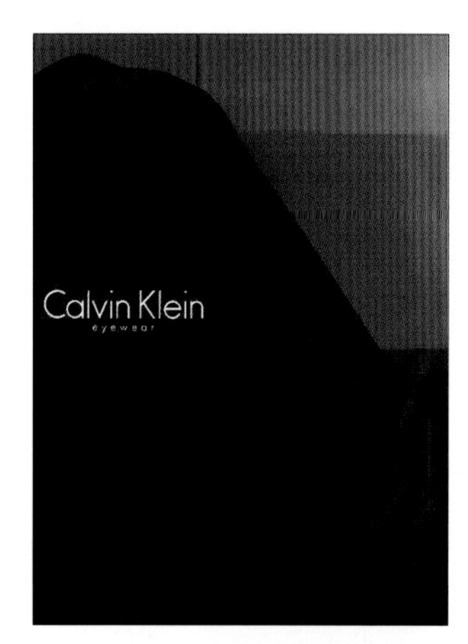

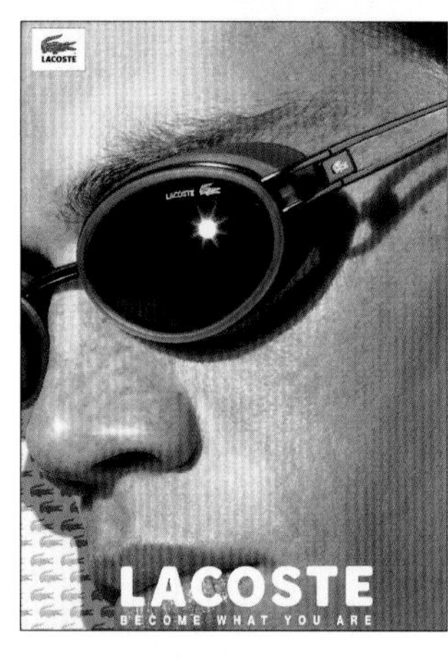

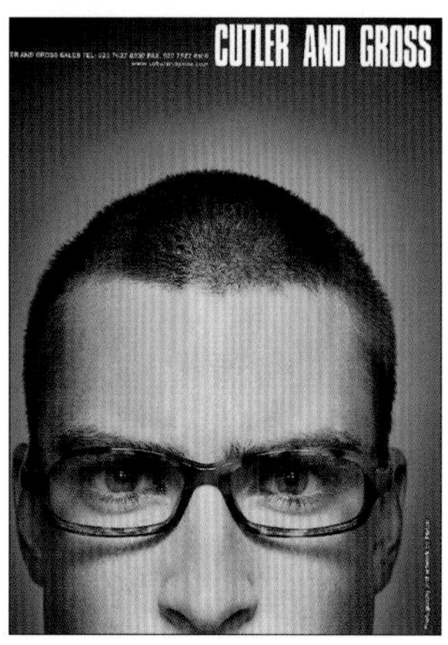

Eyewear advertised in Wallpaper magazine, April and June 2000.

Sunglasses are probably more subject to changes of fashion than other types of eyewear, yet the classic Ray-Ban seems to have been in fashion forever. Ray-Bans were developed by Baush and Lomb in the 1930s as anti-glare goggles for US Army Corps pilots exposed to intense ultraviolet light at high altitudes. The teardrop shape was devised to accommodate the field of vision of the human eye. During World War II Ray-Bans were produced exclusively for the armed forces and banned from the civilian market. During the Korean War General MacArthur, commander-in-chief of the UN forces, conspicuously sported a pair of Large Metal Sunglasses, military jargon for Ray-Bans.

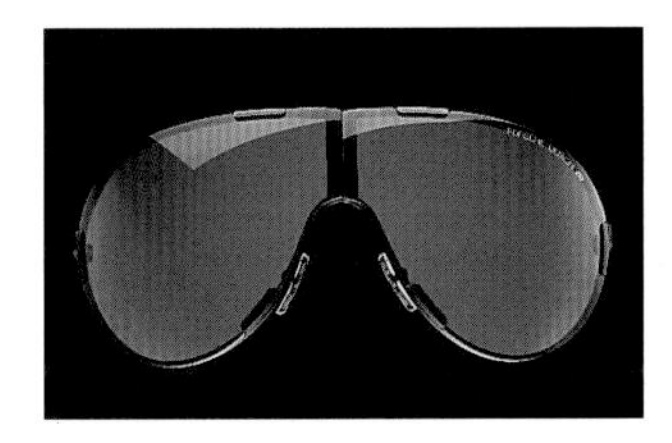

In 1986 Austrian Porsche Design added a third fold – between the lenses. Panorama Folding was manufactured by Nava of Italy.

A white mobility stick is both a symbol and a useful tool for people with impaired vision. It helps its owner to 'see' and be seen. Off-duty, it folds down to a convenient format. This model is by California Canes of the US.

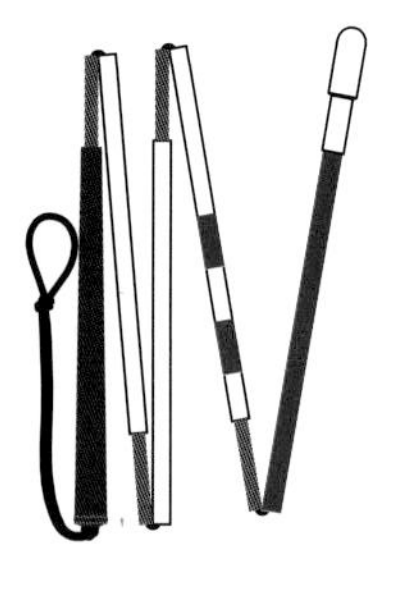

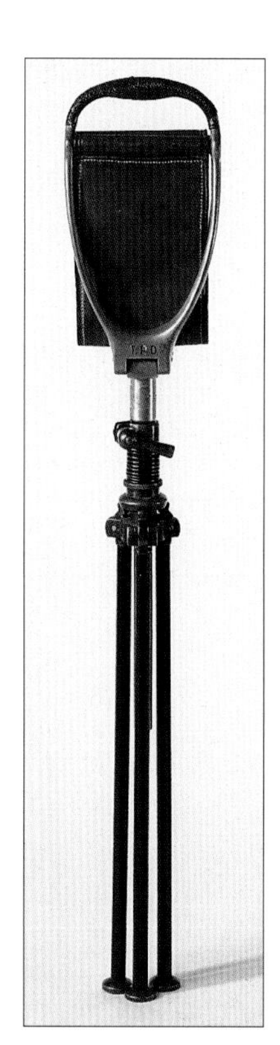

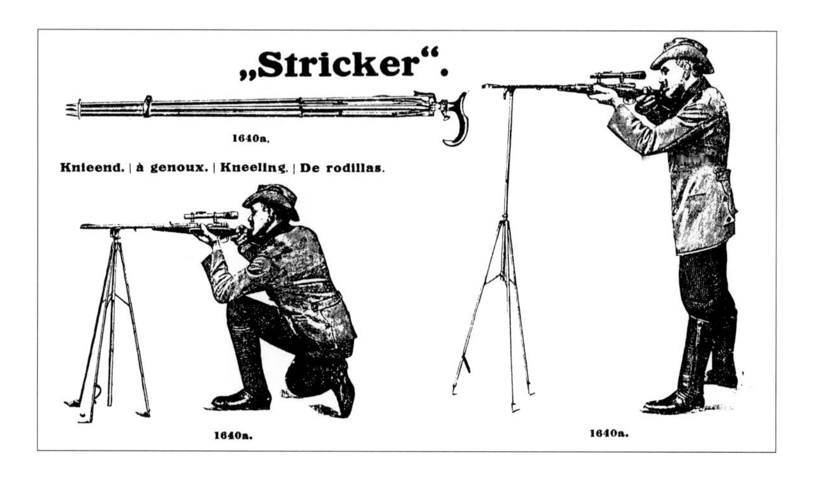

Stricker shooting tripod, from the Alfa catalogue, Germany 1918.

Portable tripod shooting seat, made with gunsmith precision

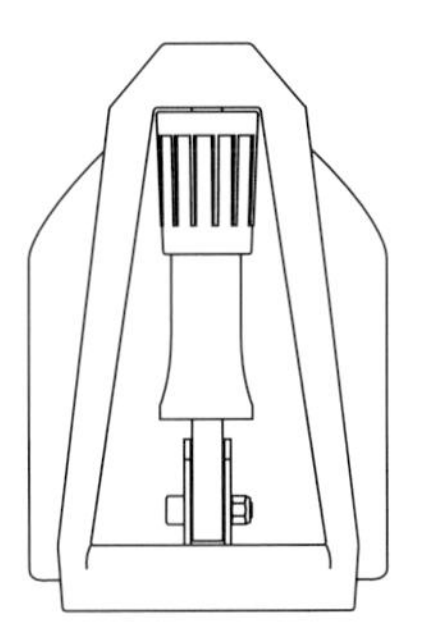

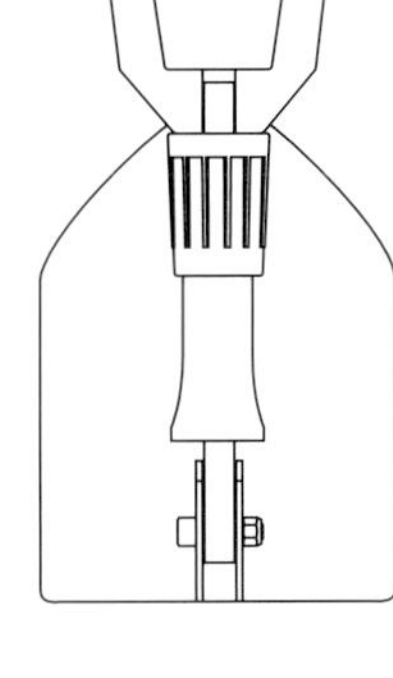

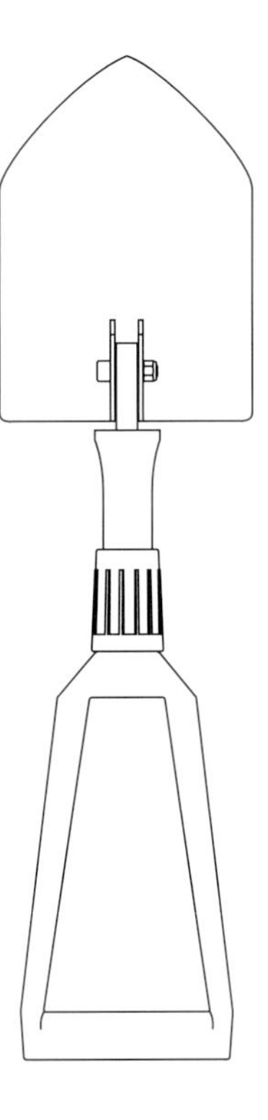

This compact earth-moving device has been designated official NATO gear. It is still called a spade. Designed by Olavi Lindén in 1995 and manufactured by Fiskars, Finland.

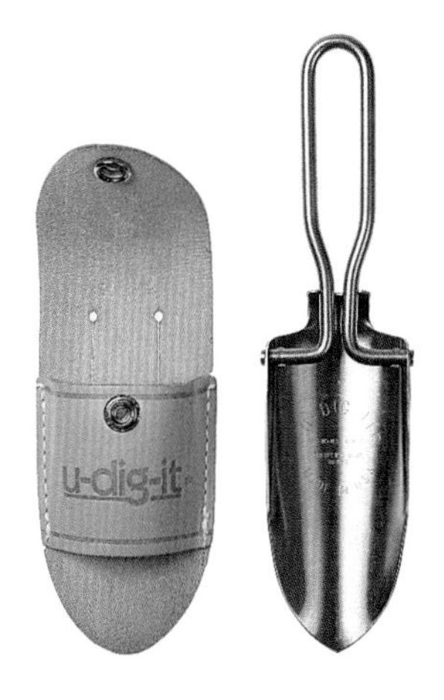

The u-dig-it folding hand shovel of stainless steel addresses the aesthetic sense as much as the practical one. Designed and manufactured by u-dig-it Enterprises, US.

Cabin lamp designed by Erik Magnussen in 1990 and manufactured by Harlang & Dreyer of Denmark. The light comes on when the shade is tipped out. It can also be turned to throw the light to one side.

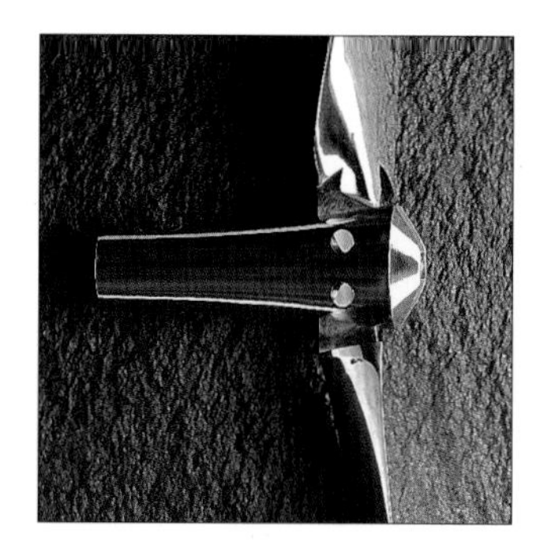

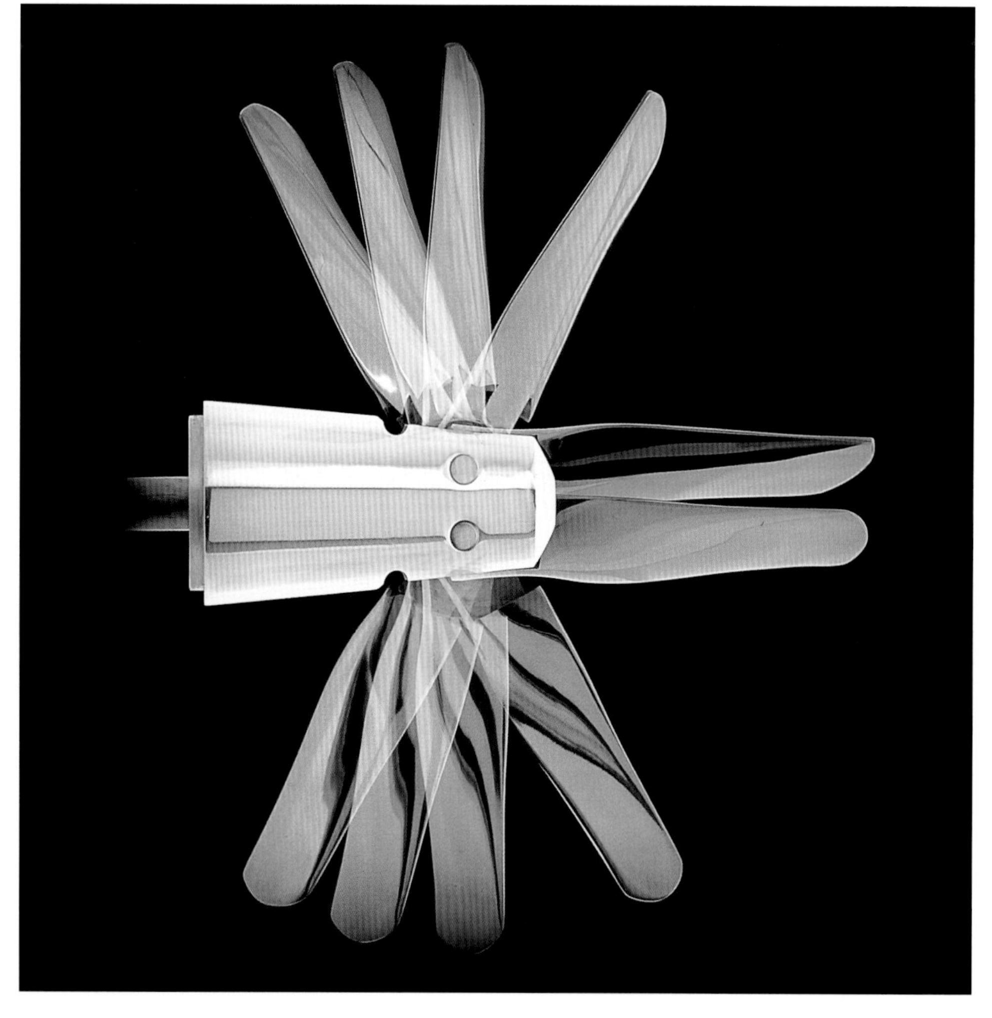

Beauty is not confined to objects made with aesthetics in mind. The 1975 folding propeller by Gori Marine of Denmark was designed to work invisibly, below the surface of the water. Nevertheless it possesses great elegance. The idea is simple: when the wind is up the propeller is automatically folded. When the weather is calm, the motor is engaged and the folding propeller automatically unfolds for use.

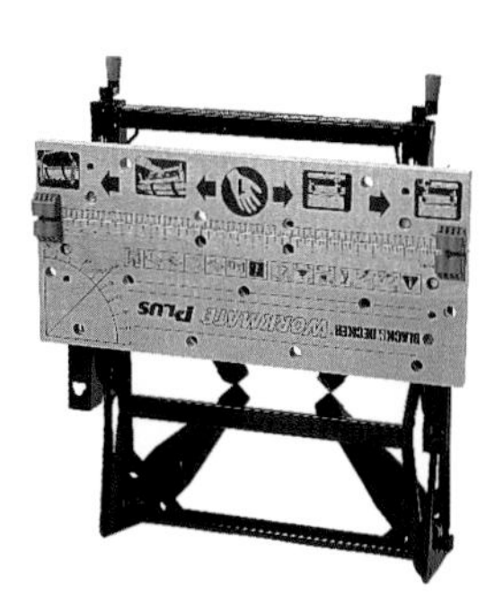

The history of the Black & Decker Workmate contains all the elements that designers-cum-entrepreneurs love to hear. When its inventor, Ron Hickman, spoiled a fine old Windsor chair while using it as a workbench, he identified a problem. This inspired him to design a collapsible workbench, with which he approached a number of manufacturers. They all turned the project down, so in 1968 Hickman set up in production himself. Four years later, in 1972, Black & Decker in the UK saw his success and revoked their rejection. By the mid-1990s more than 20 million Workmates had been sold.

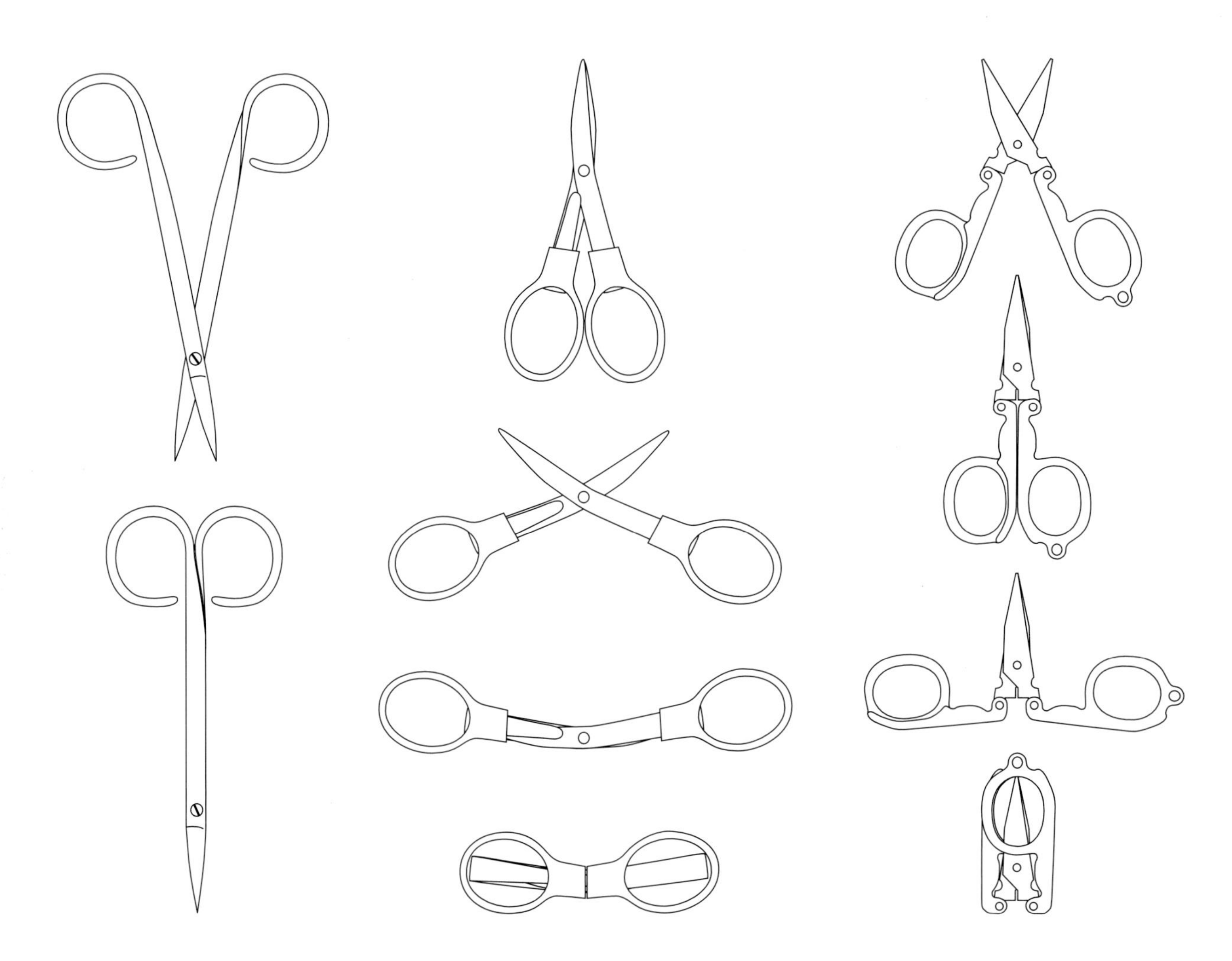

A normal pair of scissors is only a quasi collapsible, since the folded state is both active and passive.

Pocket scissors that fold once more into a shape that is purely passive are genuine collapsibles.

Travelling light is a matter of volume as well as weight, and every little helps, even collapsible hair- and toothbrushes.

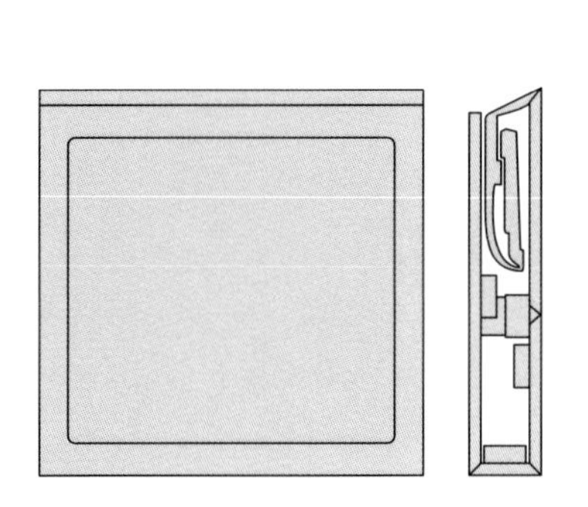

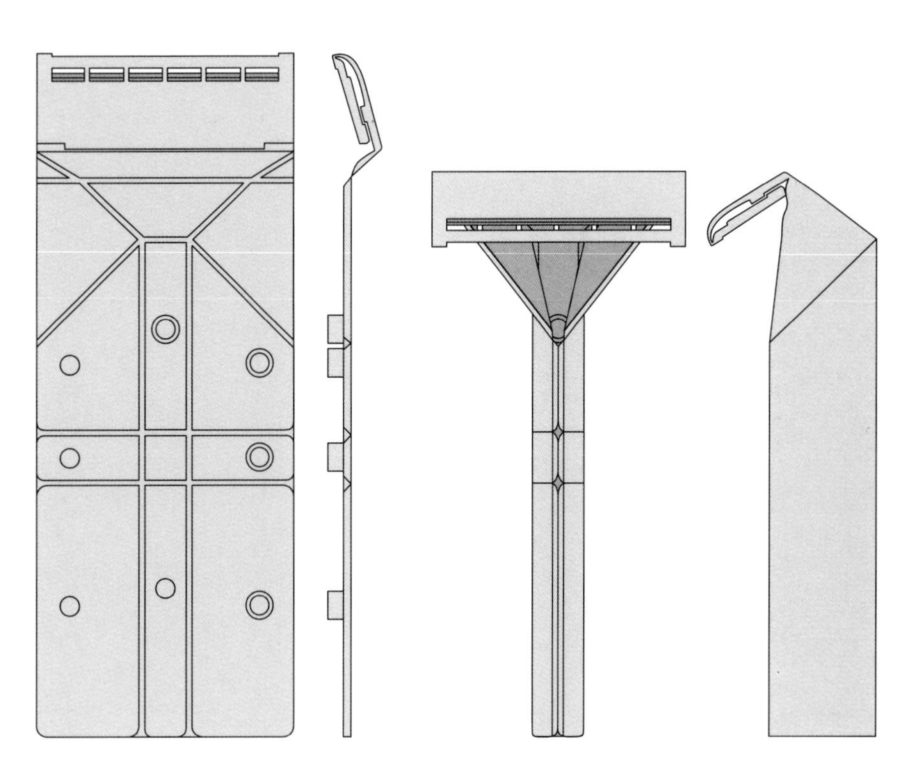

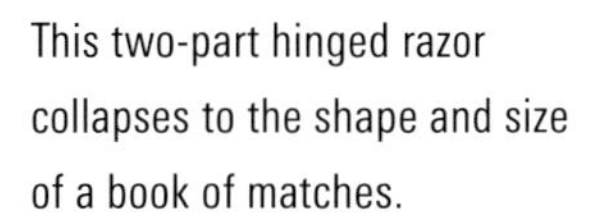

This two-part hinged razor collapses to the shape and size of a book of matches.

Travel often pulls us in opposite directions: on the one hand we need our luggage to be as small as possible; but on the other, we want to take as much as possible. The answer is collapsibles – small and convenient when not in use, fully functional when expanded.

Collapsible vehicles such as pushchairs or foldable bicycles are typically folded out during transport and collapsed before and after – the opposite of luggage, since transport is their state of use. A convertible car, on the other hand, does the opposite. Its retractable – ie collapsible – roof is typically folded down when driving in fine weather or when there are spectators around to admire, and pulled back in bad weather or when the car is not in use.

The 1999 Mercedes SLK is a sports car with an unusual retractable roof. This is not a soft top but a hinged hard roof which folds in three parts to disappear in the trunk.

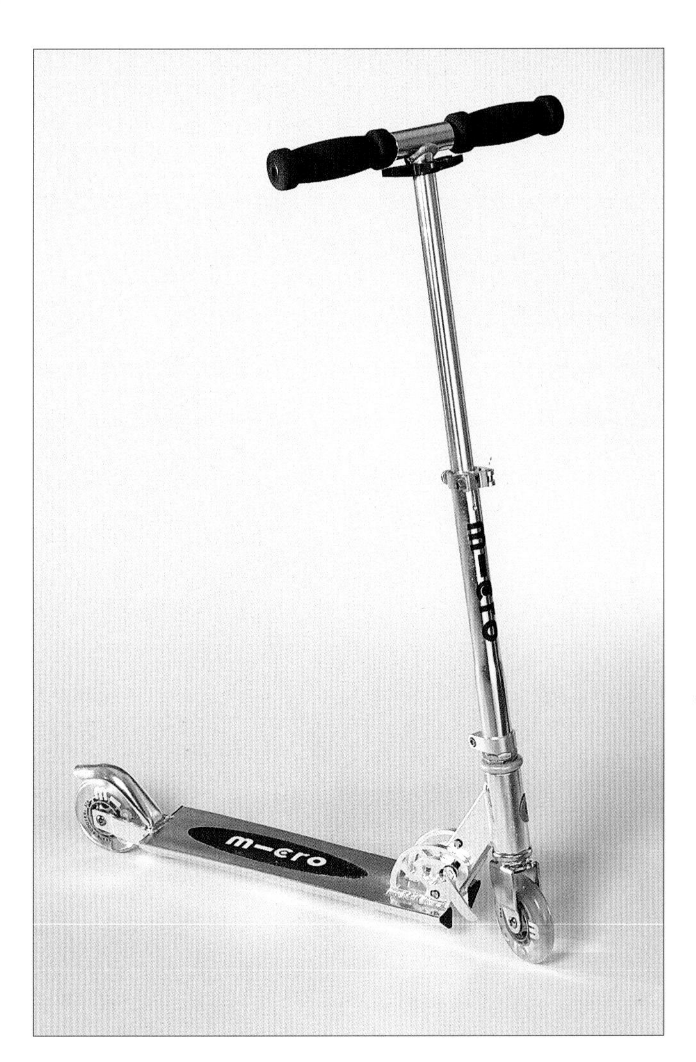

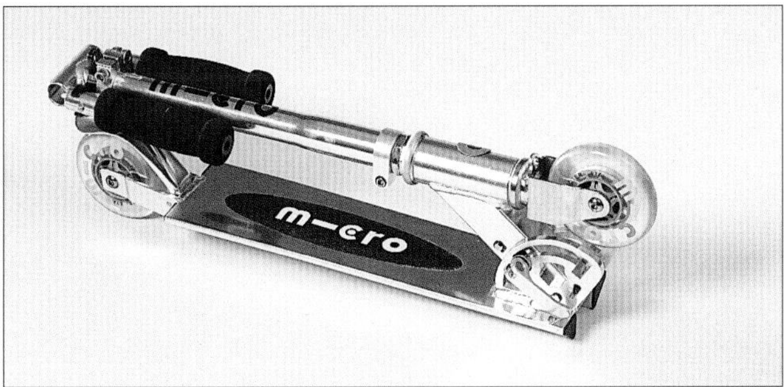

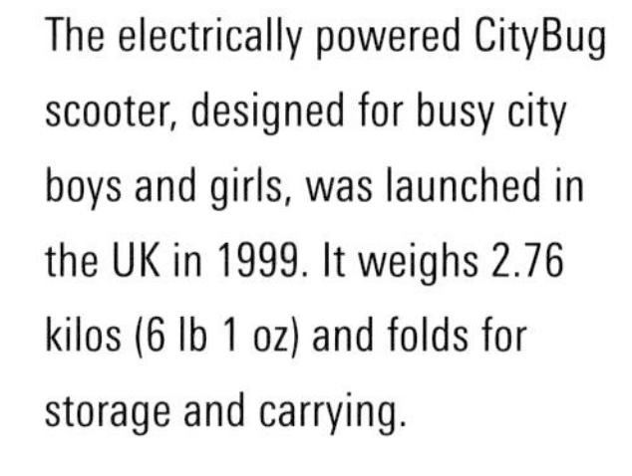

The electrically powered CityBug scooter, designed for busy city boys and girls, was launched in the UK in 1999. It weighs 2.76 kilos (6 lb 1 oz) and folds for storage and carrying.

Man's ingenuity knows no bounds when it comes to saving his own energies, and to getting around his own rules. To move long distances he invents a vehicle. To move short distances he invents a quasi-vehicle. The Micro skate-scooter from CityBug looks innocent enough to venture into any pedestrian zone.

BROMPTON

◀ ▲
The first Brompton Bicycle was designed in 1975 by landscape gardener Andrew Ritchie, in his bedroom overlooking London's Brompton Oratory.

This bicycle trailer designed by Erik Magnussen in 1985 folds flat and runs parallel with the bicycle when not in use.

War is all about change, and change creates the need for new collapsibles. The Wellbike foldable motorcycle was designed in Britain in the 1940s.

Another bit of British World War II ingenuity: a foldable military bicycle designed in the 1940s.

Solving a problem in one field using lessons learned in another is one of the great secrets of successful design. When British aeronautical engineer Owen Maclaren retired, he embarked upon a second career capitalizing heavily on his insights into light but strong supporting structures. He used his knowledge to design the Maclaren baby buggy, a lightweight aluminium pushchair on softly rolling balloon tyres. He made the first prototype design in 1965, and the earliest model went on the market two years later. Rather than folding flat these buggies fold tall and thin for easy storage. They gave birth to a company now manufacturing more than half a million buggies per annum.

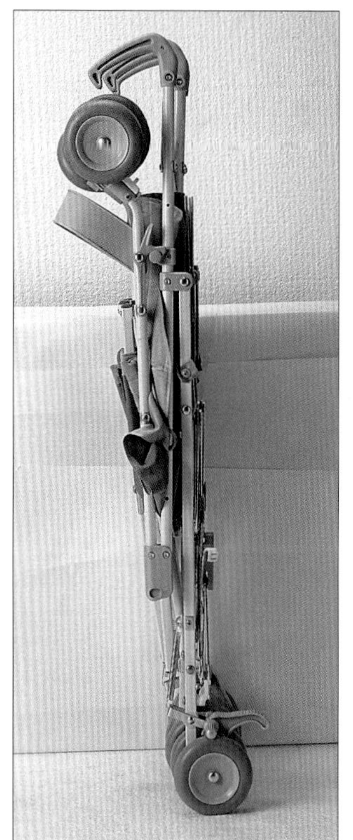

The Davis Collapsible Go-Cart, an early answer to an old need, from the Sears, Roebuck & Co. catalogue of 1908.

The 1954 RNAF Gannet, by Fairey in the UK, is a three-seat shipboard early-warning aircraft that was designed to take off and land from an aircraft carrier. Its wings fold back for compact on-board storage.

The aircraft carrier USS Franklin D. Roosevelt Sentinel with collapsible aircraft on the flight deck, *c*. 1970. These days many helicopters and military aircraft have rotor blades and wings that fold to save space on aircraft carriers.

Knifeless man is lifeless man, or so the saying goes. The knife was man's first and most important tool. It helped primitive humans to meet some of their most elementary needs: gathering food, building shelter, and keeping enemies – animal or human – at bay. The knife has been with us for thousands of years, as has the folding knife. Archaeological finds at Herculaneum and Pompeii demonstrate that folding knives have served us for at least two millennia. Pocket knives are more recent, but only because pockets are not that old. The evolution of knives reflects the evolution of tools in general. They have been created to serve three generic purposes: survival, convenience and pleasure – in that order. The basic knife, a blade mounted on a handle, is a survival tool. Folding knives, which may have one or more blades, usually cater to the second and third purposes. The idea of a folding knife is double. It is smaller to carry when folded, and of course safer, since the cutting edge can't do accidental harm. It may also be a sheer pleasure to own and use. Since the mid-18th century most folding knives have had back springs made from spring steel, which was invented in Sheffield in 1742. The spring keeps one or more blades in position whether the knife is open or closed. Folding knives without a back spring came to be called penny knives, since they were cheaper. Armed forces across the world have commissioned innumerable specialized knives of ingenious design and multiple uses, all of them now collector's items. This stainless steel British naval officer's knife from 1955, manufactured by J.H. Thompson, has a blade, can opener, bottle opener and spike. The King's Mark – the arrow on the handle – signals that it is government property.

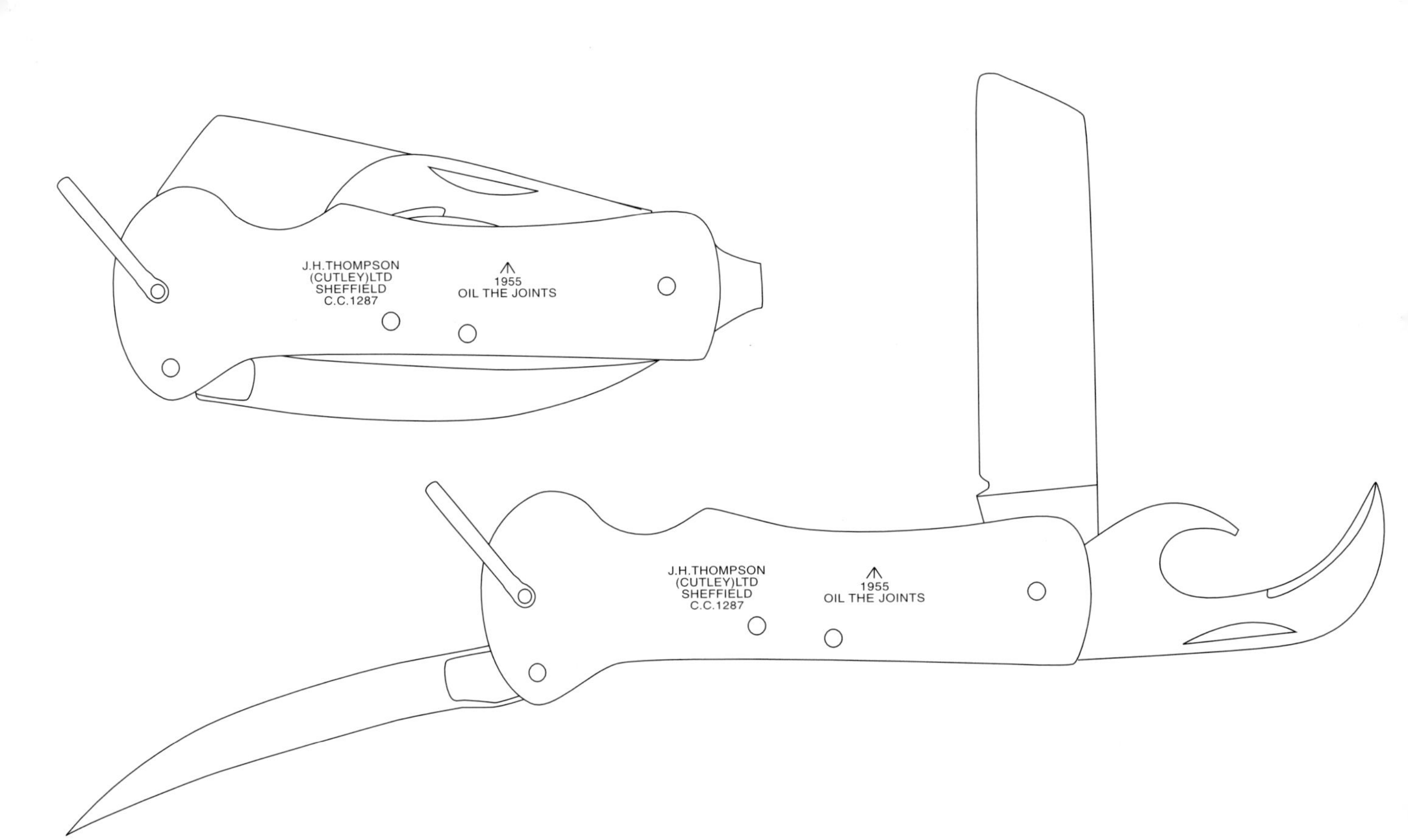

Folding knives may have multiple blades and any number of more or less useful extra implements. Corkscrews were added in the later 19th century, followed by other implements such as a cartridge remover, screwdriver and can opener. The 1907 Army & Navy Stores catalogue featured more than a hundred different pocket knives.

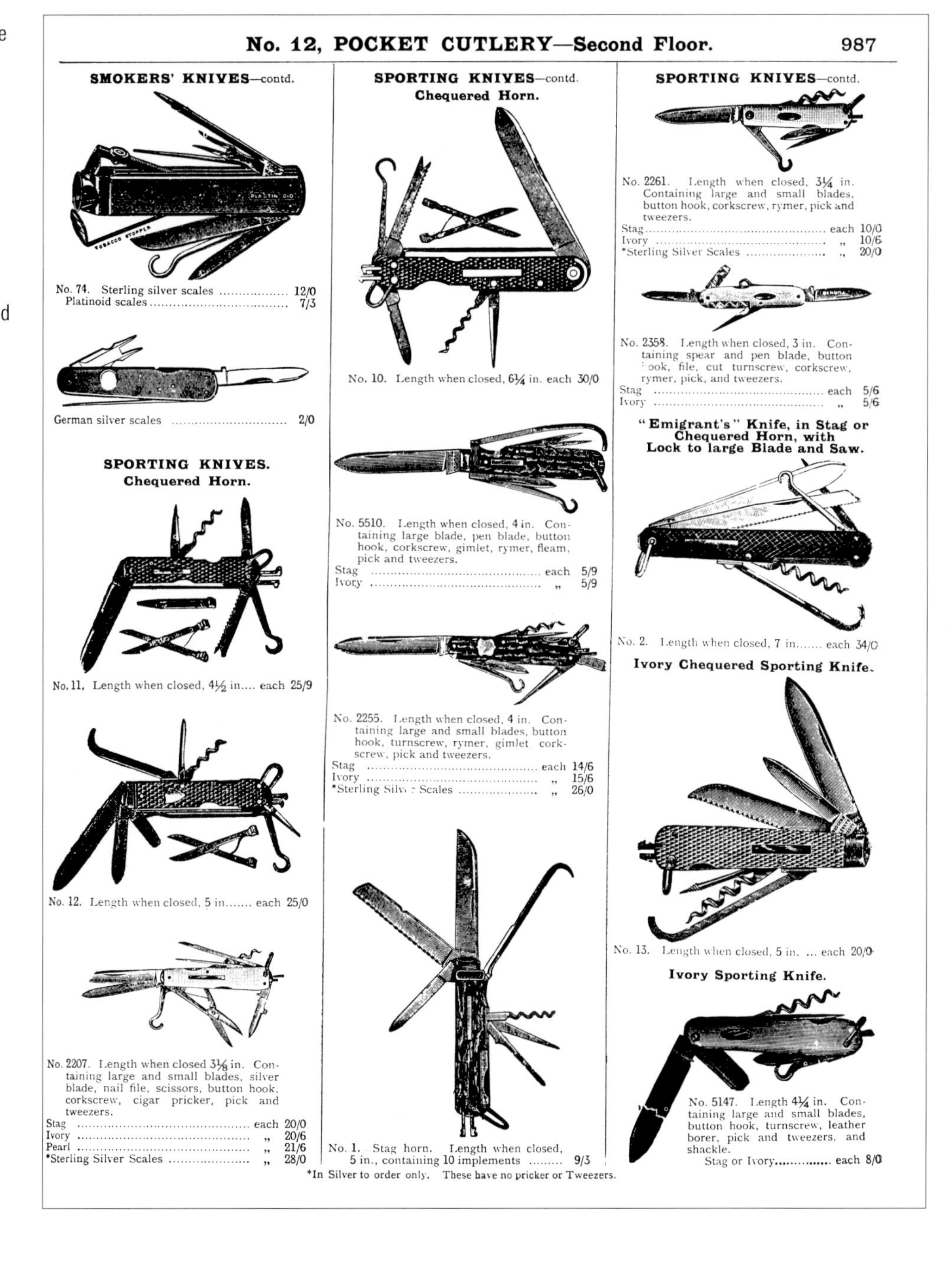

No. 12, POCKET CUTLERY—Second Floor. 987

SMOKERS' KNIVES—contd.

No. 74. Sterling silver scales 12/0
Platinoid scales 7/3

German silver scales 2/0

SPORTING KNIVES.
Chequered Horn.

No. 11. Length when closed, 4½ in.... each 25/9

No. 12. Length when closed, 5 in....... each 25/0

No. 2207. Length when closed 3⅛ in. Containing large and small blades, silver blade, nail file, scissors, button hook, corkscrew, cigar pricker, pick and tweezers.
Stag .... each 20/0
Ivory .... „ 20/6
Pearl .... „ 21/6
*Sterling Silver Scales .... „ 28/0

SPORTING KNIVES—contd.
Chequered Horn.

No. 10. Length when closed, 6¼ in. each 30/0

No. 5510. Length when closed, 4 in. Containing large blade, pen blade, button hook, corkscrew, gimlet, rymer, fleam, pick and tweezers.
Stag .... each 5/9
Ivory .... „ 5/9

No. 2255. Length when closed, 4 in. Containing large and small blades, button hook, turnscrew, rymer, gimlet corkscrew, pick and tweezers.
Stag .... each 14/6
Ivory .... „ 15/6
*Sterling Silver Scales .... „ 26/0

No. 1. Stag horn. Length when closed, 5 in., containing 10 implements 9/3

SPORTING KNIVES—contd.

No. 2261. Length when closed, 3¼ in. Containing large and small blades, button hook, corkscrew, rymer, pick and tweezers.
Stag .... each 10/0
Ivory .... „ 10/6
*Sterling Silver Scales .... „ 20/0

No. 2358. Length when closed, 3 in. Containing spear and pen blade, button hook, file, cut turnscrew, corkscrew, rymer, pick, and tweezers.
Stag .... each 5/6
Ivory .... „ 5/6

"Emigrant's" Knife, in Stag or Chequered Horn, with Lock to large Blade and Saw.

No. 2. Length when closed, 7 in....... each 34/0

Ivory Chequered Sporting Knife.

No. 13. Length when closed, 5 in. ... each 20/0

Ivory Sporting Knife.

No. 5147. Length 4¼ in. Containing large and small blades, button hook, turnscrew, leather borer, pick and tweezers, and shackle.
Stag or Ivory.............. each 8/0

*In Silver to order only. These have no pricker or Tweezers.

Opinel knives were designed by Joseph Opinel in Savoie, France, in 1890, and have remained virtually unchanged ever since. They are single-blade folding knives without a back spring. In all except the smallest, the open blade is locked in place by a twist-ring. Beechwood handle and blade are both embellished with Opinel's logo, *la Main couronnée*.

The stamp on the handle is black when the blade is of carbon steel and red when it is of stainless steel. The standard Opinel comes in thirteen sizes, numbered 1 to 13. Opinel was – and still is – a peasant knife, but its vernacular and almost timeless no-nonsense design has turned cosmopolitan. Today it is found in museums and sold throughout the world. It also comes in various colours and in several special-purpose models.

Barbers, waiters, and many other professionals use their own special folding knife. The razor knife – this one produced by C.V. Heljestrand of Sweden – is a lethal collapsible if not treated with great respect, and the blade always folded into the handle when not in use. One Danish king was so scared of knives that he had his beard burned off.

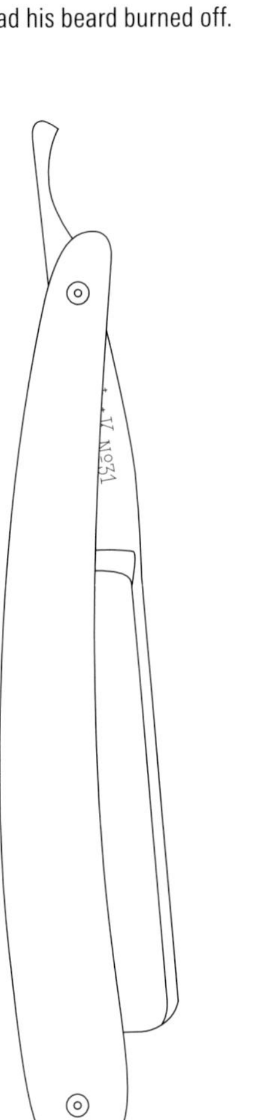

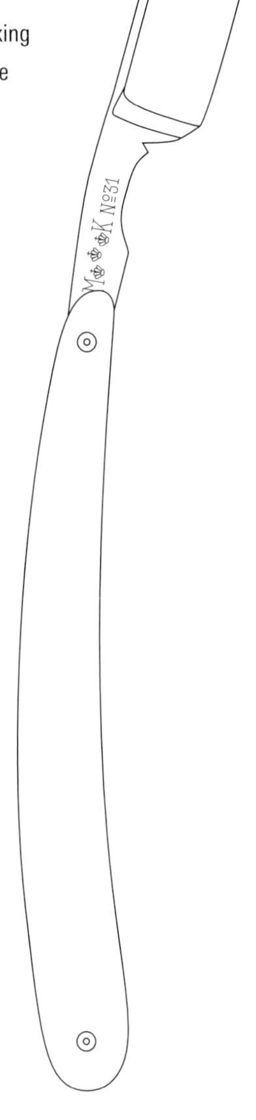

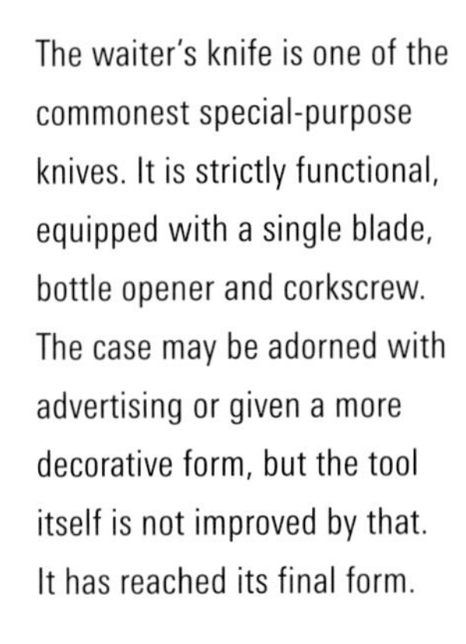

The waiter's knife is one of the commonest special-purpose knives. It is strictly functional, equipped with a single blade, bottle opener and corkscrew. The case may be adorned with advertising or given a more decorative form, but the tool itself is not improved by that. It has reached its final form.

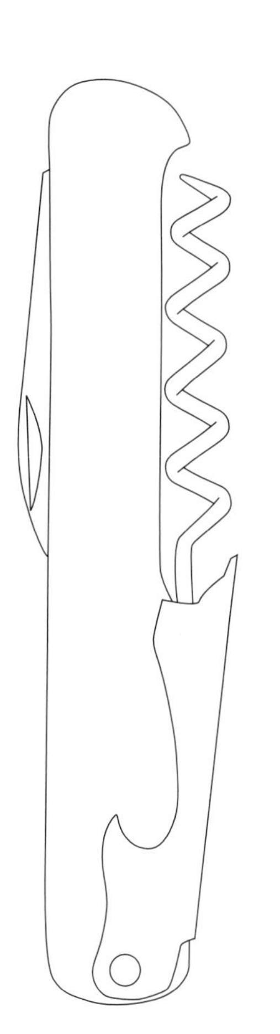

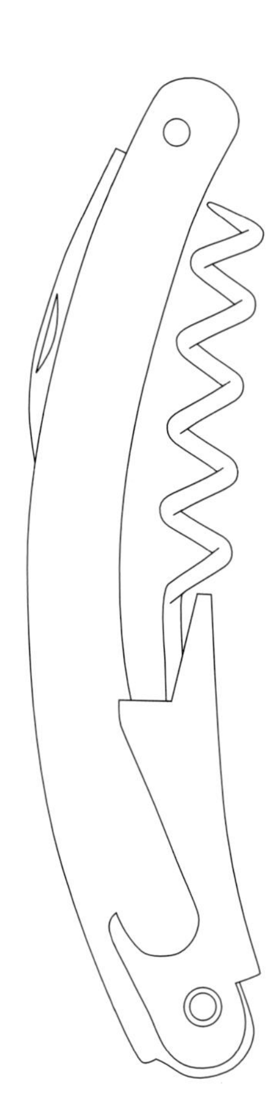

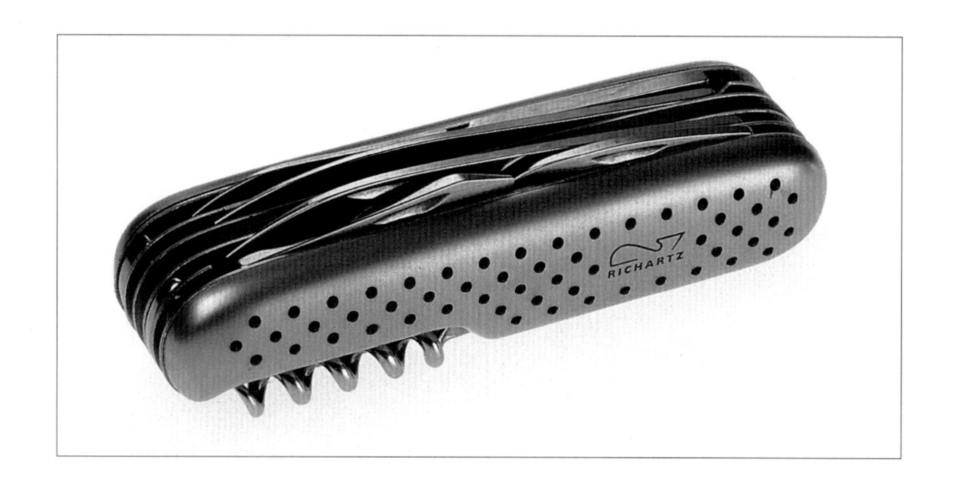

The Structura seven-piece folding knife by Richartz of Germany is a *slotknife* with pull-out cutlery – a feature as likely to provide the owner with conversation as food. The surface of the stainless steel handle is dotted with small rubber studs that provide grip whatever the knife is used for.

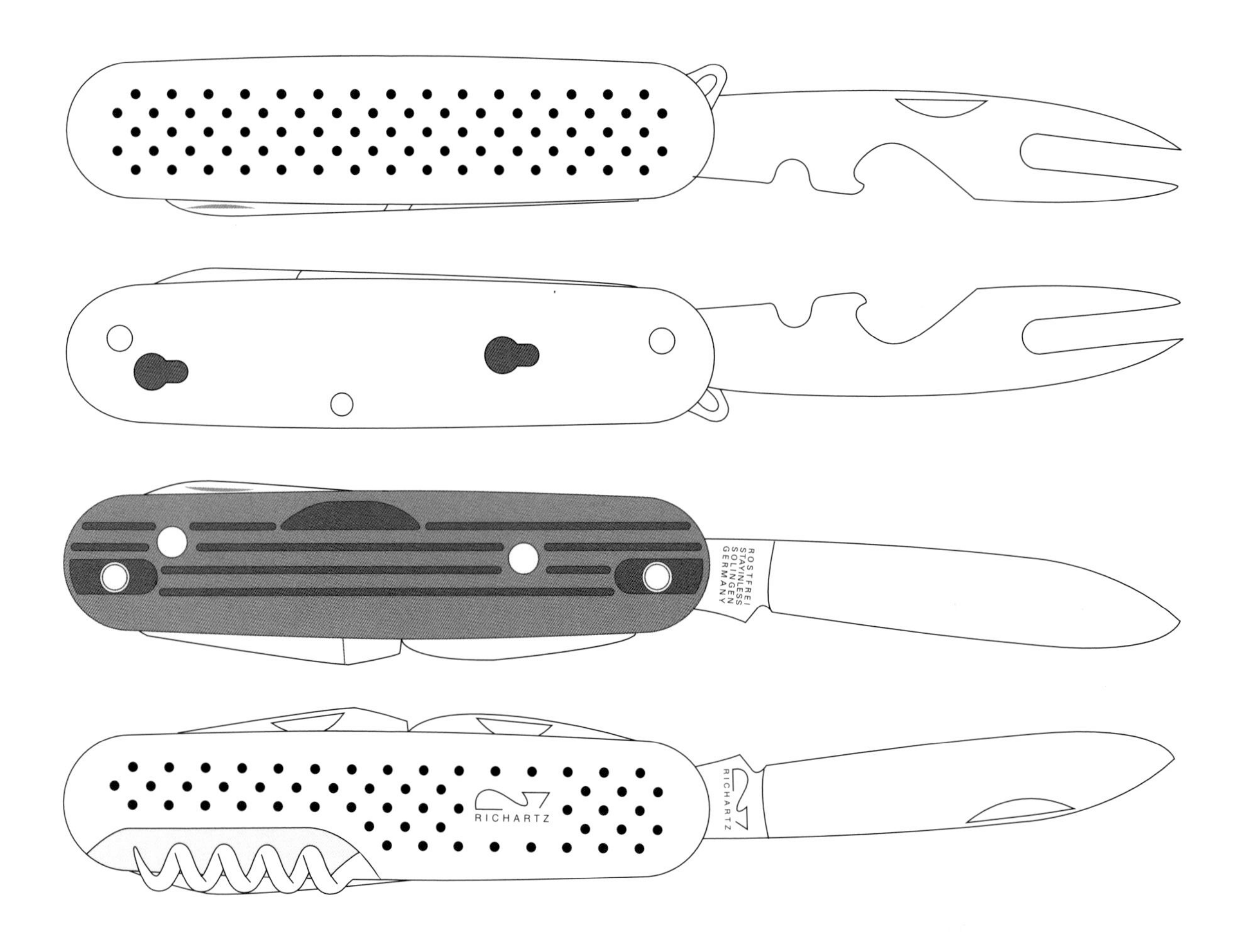

Most pocket knives have their blades mounted to disappear into one side of the handle when not in use. Butterfly – or double handled – knives have handles that divide into two halves and fold around the retracted blade. Whether the butterfly knife is open or closed, the two halves lock together to form a unity, turning back to back when the blade is out. Butterfly knives originally came from the Philippines, where they are called *balisong.* Because of their violent potential they are prohibited in some countries.

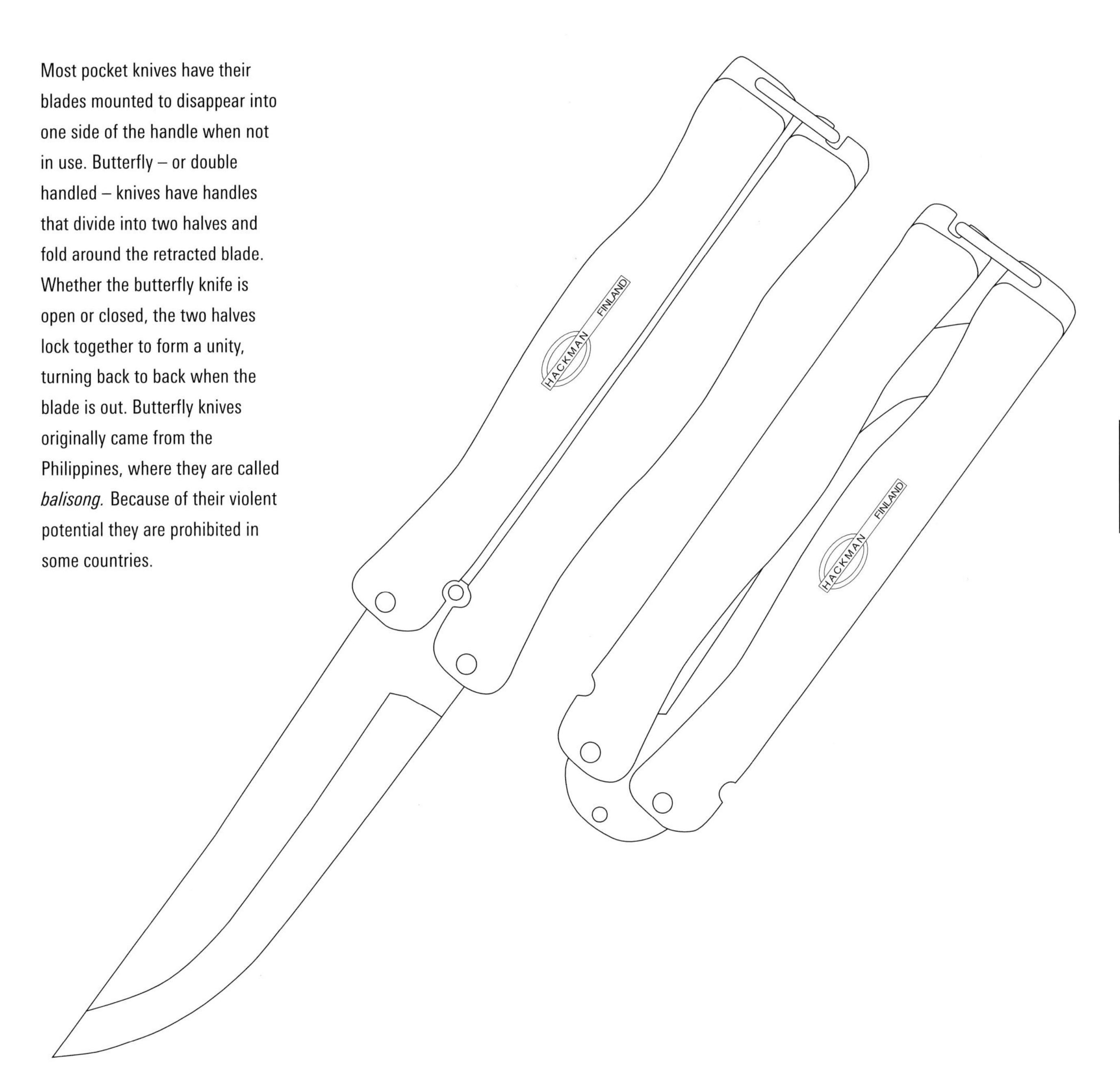

The term Swiss Army Knife covers a family of knives made by two different Swiss companies: Carl Elsener and Wenger. Carl Elsener also uses the more familiar name Victorinox, a contraction of the name of the founder's wife, 'Victoria', and 'inoxidable', French for stainless (steel). In 1893 Carl Elsener won a contract to produce a standard knife for the Swiss army equipped with a blade, spike, can-opener, and screwdriver. This knife had wooden handles. In 1897 Carl Elsener developed and patented the Officer's Knife which had an additional small blade and corkscrew and the characteristic red handle. This type – in many varieties – has become known as the Swiss Army Knife.

Cyber Tool from Carl Elsener is a recent development marketed as 'the pocket knife for the third millennium'. Cyber Tool has 34 different gadgets, many of them for the computer nerd who is into hardware as well as software.

Since 1961, the regulation knife of the the Swiss army's rank and file has been a four-piece model with six tools: blade, can opener, small screwdriver, spike, bottle opener, and larger screwdriver. The knife has aluminium handles – red until 1965 – adorned with the Swiss flag in the shape of a shield. It is the most beautiful and functional Swiss Army Knife.

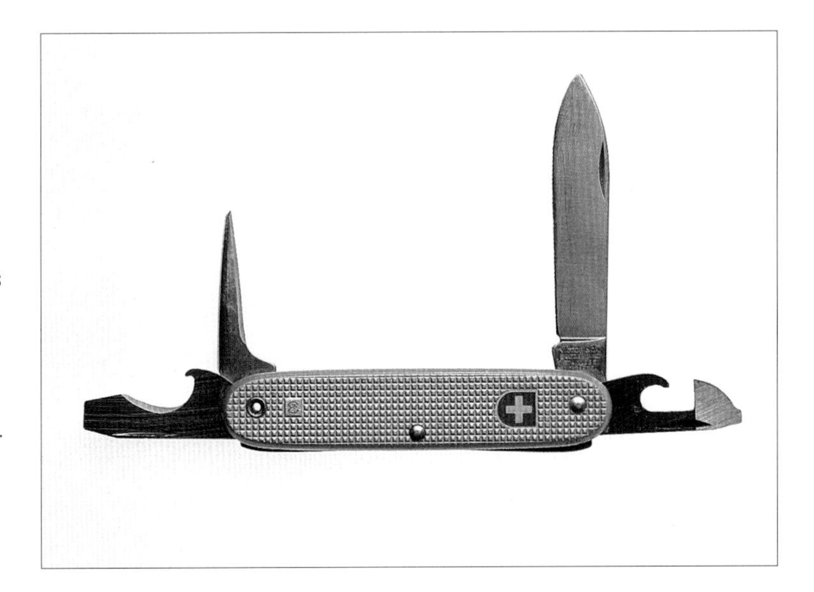

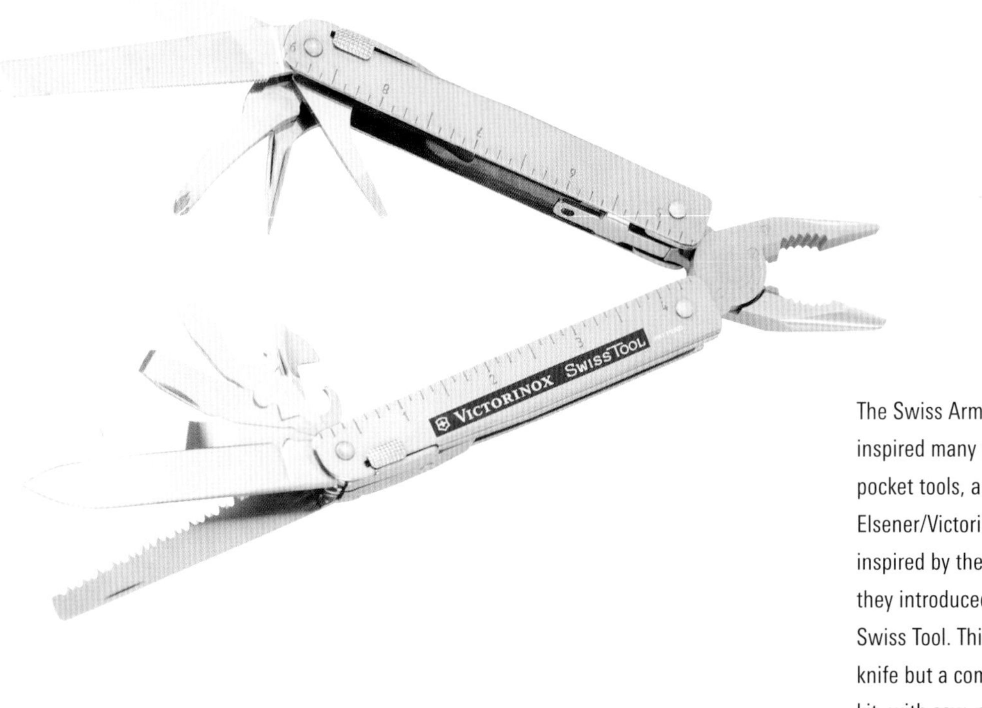

The Swiss Army Knife has inspired many manufacturers of pocket tools, and maybe Carl Elsener/Victorinox were in turn inspired by their colleagues when they introduced the Victorinox Swiss Tool. This is not a pocket knife but a complete pocket tool kit, with saw, chisel, ruler, screwdriver and 20 other implements.

The Leatherman Supertool is only the latest stage in a development which began when a corkscrew was first added to a folding knife. Numerous gadgets and devices have been installed since alongside the basic blade, but Leatherman, Gerber and SOG tools are in a different league. They are too heavy for a normal pocket and have to be kept in special clips or in heavy-duty sheaths of leather or equally strong materials.

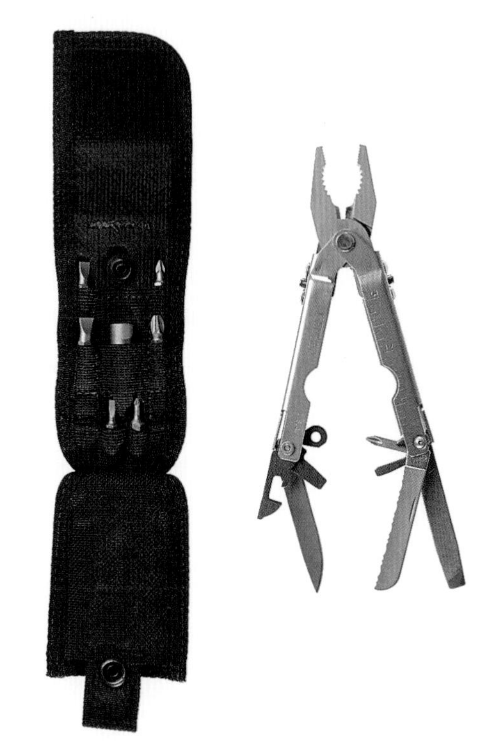

The Multi-Plier pocket tool kit includes a range of screwdriver tips. Manufactured by Gerber in the US.

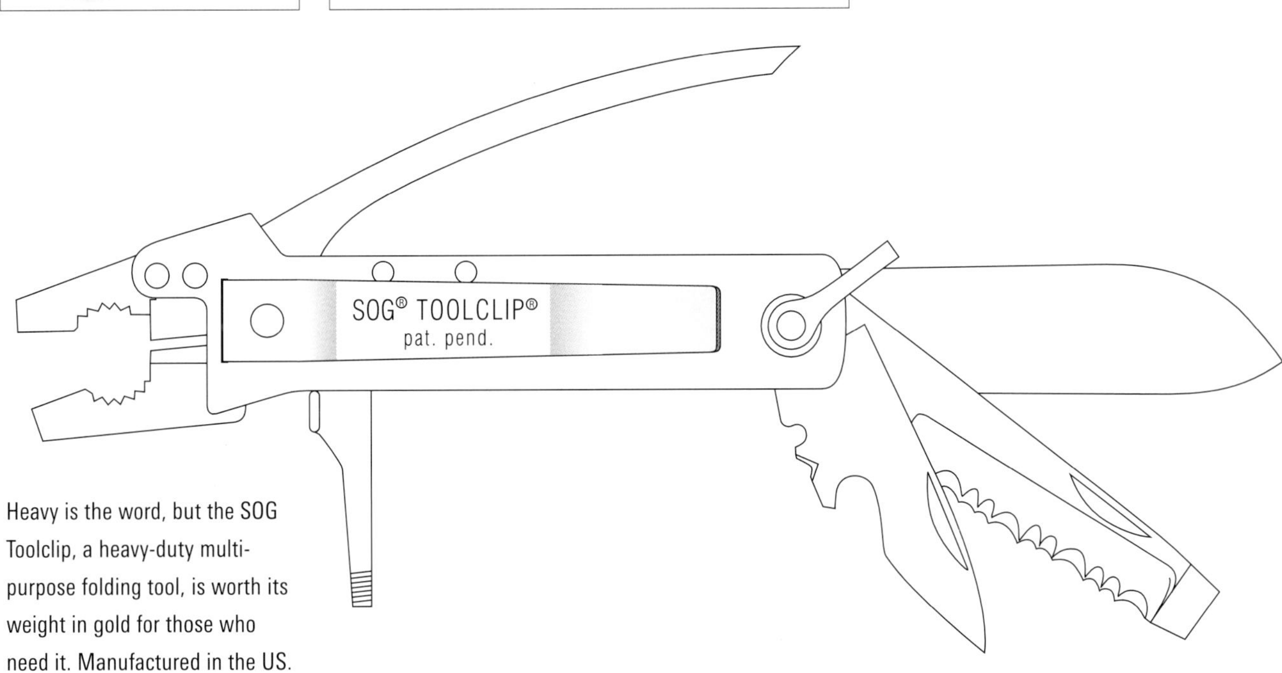

Heavy is the word, but the SOG Toolclip, a heavy-duty multi-purpose folding tool, is worth its weight in gold for those who need it. Manufactured in the US.

Rolling

The newspaper stick holds the paper like an old-fashioned clothes peg. When not in use the newspaper can be rolled around the stick and hung on the wall by the hook at its tip.

Collapsibles are objects that are folded and unfolded again and again. That rules out reels of yarn but rules in reels of fishing line or hoses, whether they serve to extinguish a fire or to water a garden. Electrical extension cords and other supply lines are also rolled and unrolled repeatedly, as are tape measures and dog leads.

Rolls of papyrus were the forerunners of our folded and bound paper books. Architects' drawings and topographical maps are still made on large rolled sheets of paper. Sheets of cloth and other materials may be repeatedly rolled up for later unrolling. Other examples include rollerblinds, temporary film screens, maps, charts and photographers' backdrops.

Fine Panama hats may be rolled when not worn. In fact it is an accepted test of quality that the very best Panamas roll small enough to pass through a wedding ring. Some of the headgear from Italian milliner Borsalino can be rolled too, and is sometimes packaged and delivered in tubular cans. In the circus, the stablemen roll out the carpet for the unicycling acrobats and roll it up again before the Cossacks on horseback enter the ring.

Some tool and repair kits are stored in cloth sheets lined with pockets that roll up for protection and roll out for display. Housemaids traditionally used the same method to store silver cutlery. Some ceremonies – military or other – involve the ritual unfurling of rolled banners.

Many cafés and library reading rooms roll the daily newspapers on sticks, an arrangement that keeps the papers in order and limits their untimely removal from the premises.

Japanese women examining a scroll. *Celebrated Beauties Compared to the Chushingura*, Utamaro, *c*. 1800.

Electric extension cable on a reel that winds it back in for storage.

French bulldog Gaston leading his mistress by an automatic roll-up leash.

Ambassadeur Mörrum 5600C reel for 150 m (492 ft) of 0.35 mm ($^{1}/_{100}$ in) nylon fishing line. Note the royal warrant: this reel was made by Abu Garcia, supplier of fishing gear to the Swedish royal court since 1950.

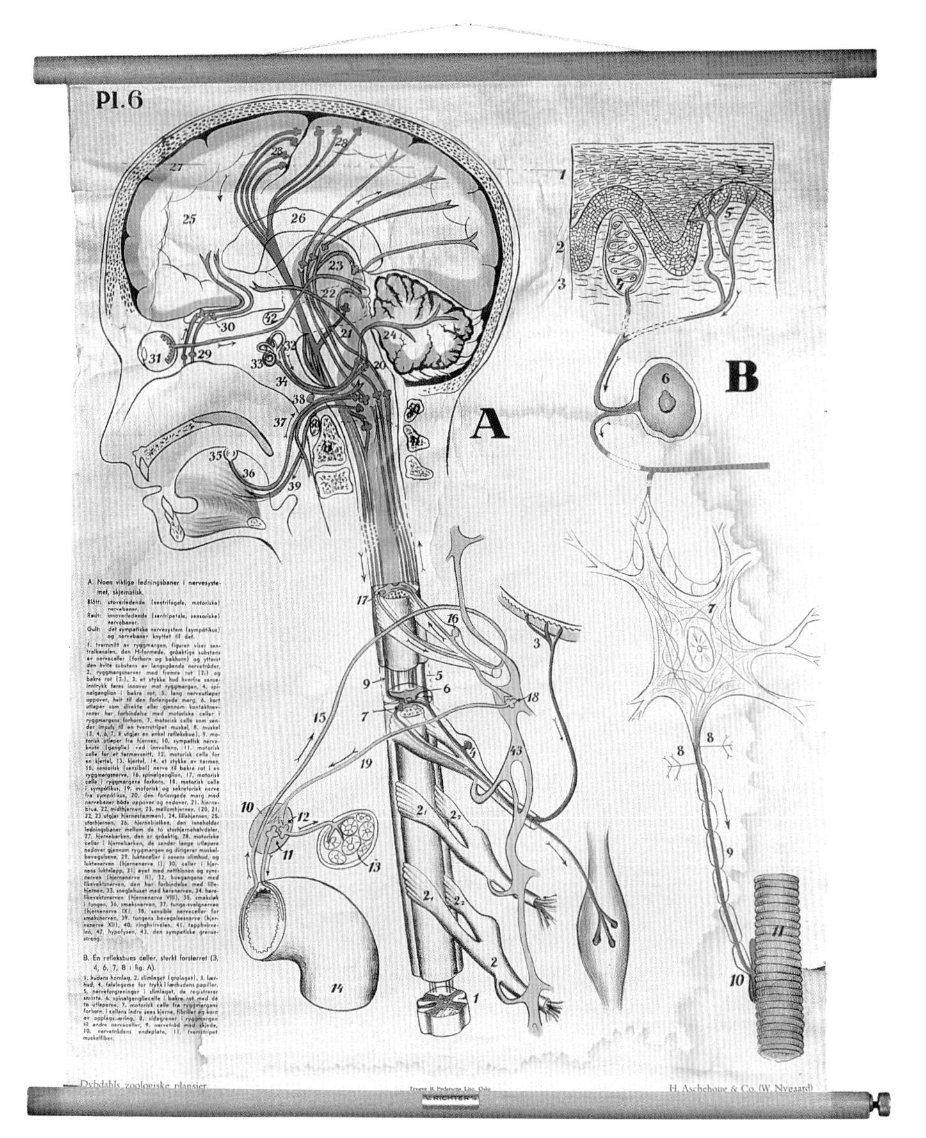

A didactic roll up, familiar to the pre-digital generation.

Early roller blinds did not have built-in springs, which were not invented until 1770. Before then, a variety of different systems were used, not all of them effective.

Panama is a country and a hat. Paradoxically, the hat does not originate in the country. It is so named because it was originally shipped to Europe from Panamanian ports, but genuine Panamas come from Ecuador. They are woven from sulphur-bleached toquilla straw, narrow strips from the leaves of the *Carludovica palmata*. The quality of a Panama is judged by its colour, shape and above all the fineness of its weaving. The more fibres per inch, the better the hat. Finest are Montecristi Superfinos and Montecristi Finos, named after their place of origin. A single Superfino may take several months to make. The Cuenca, another fine Panama named after its place of manufacture, can be woven in a week or two. It is an established test of quality that a really fine Panama will roll up small enough to pass through a wedding ring: collapsibility as a measure of quality. Panamas are traditionally delivered folded in boxes of balsa wood.

Winston S. Churchill with a Panama and a Havana.

A classic Italian Borsalino hat delivered rolled in a can.

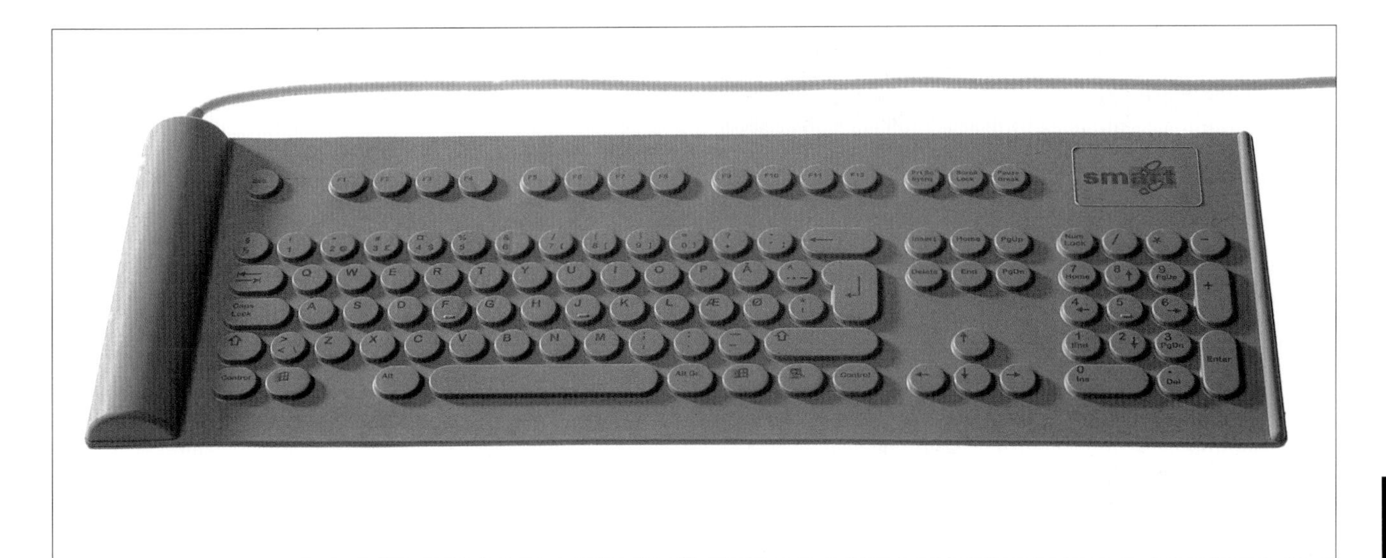

The Smart Fun roll-up rubber keyboard will withstand all the coffee, Coke and Big Mac that a nerd may spill. It can be cleaned in hot water at 60° C (140° F). Its relatives, the Smart Indupact and Smart Medic, were designed for workshops, laboratories and clinical environments, and can take even tougher cleaning. One day – or so this author foresees – all portable PCs will have ultraflat collapsible keyboards and screens that project onto the nearest wall or float freely in the air.

The Flexi-cal pocket calculator, made in China.

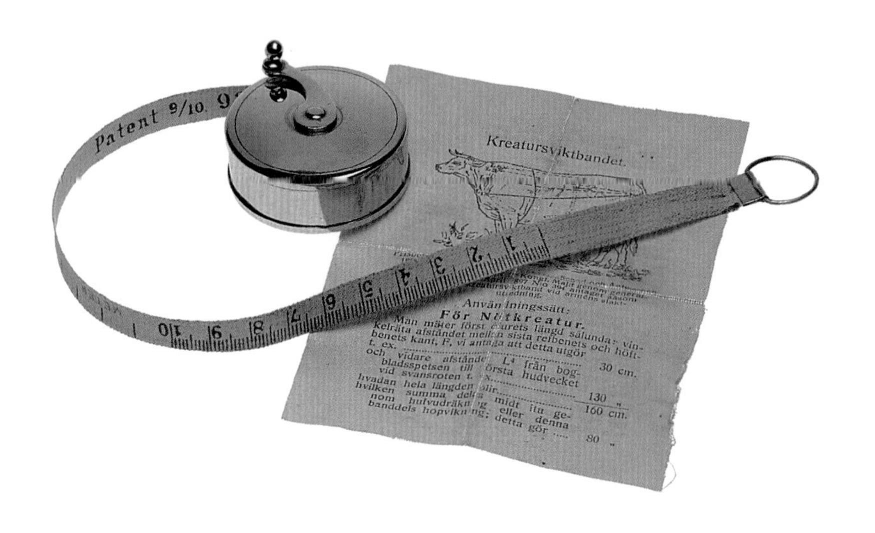

Real farmers know the time simply by lifting a cow's udder. And they can find out her weight by measuring her stomach with this retractable livestock weight-measuring tape.

The Climax lawn tennis measuring tape, from the Army & Navy Stores catalogue of 1907.

General-purpose retractable measuring tapes.

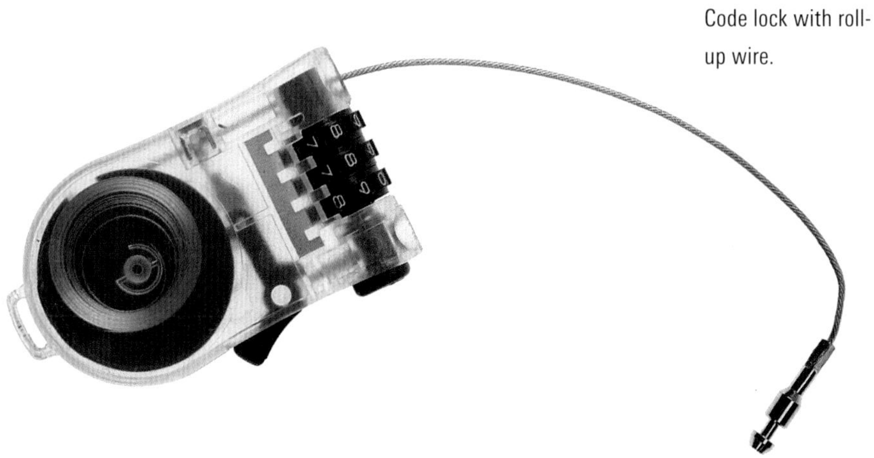

Code lock with roll-up wire.

Free-standing pine room partition that rolls out for instant privacy. It was designed by Alvar Aalto in 1935–36 and produced by Artek of Finland.

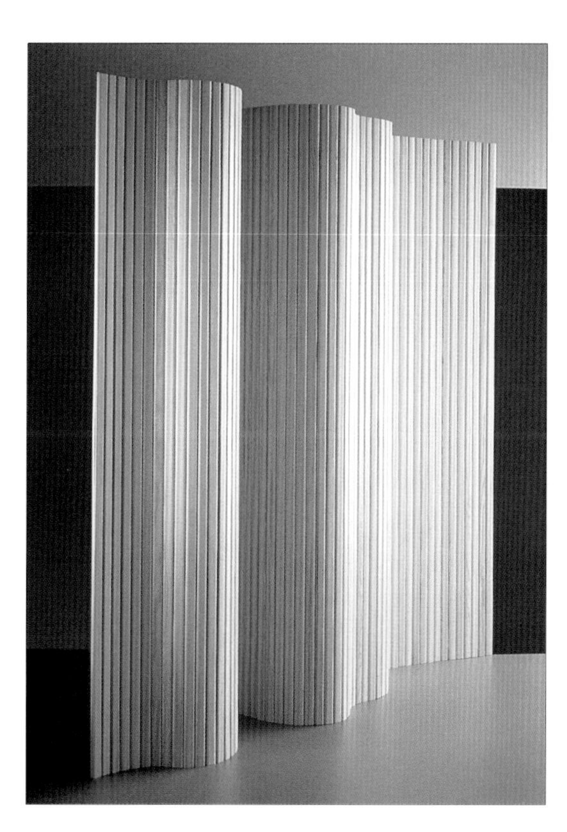

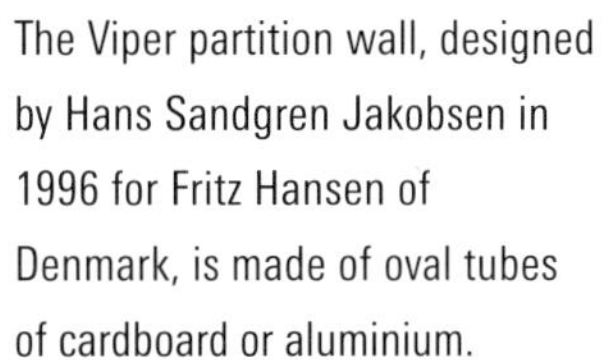

The Viper partition wall, designed by Hans Sandgren Jakobsen in 1996 for Fritz Hansen of Denmark, is made of oval tubes of cardboard or aluminium.

Sliding

Danish comedian, entertainer and world-class pianist Victor Borge on stage with his microphone. Like any stand-up comedian worth the name, Borge could fill the first five minutes of his show just by telescoping the mike.

Some collapsibles expand and contract as their parts slide open or closed: think of a paper cutter or the (now obsolete) slide rule. But the most versatile and spectacular application of the sliding principle is the telescope. In the optical telescope, a number of tubes of declining size slide in and out of one another, allowing the lens to be brought into focus and the whole to be folded down for storage; other applications of the telescope sliding principle have a similar collection of diminishing parts, not necessarily tube-shaped, which slide or twist into each other.

The old-fashioned telescope has been succeeded by less space-greedy optical devices, but its mechanical principle remains widely used in radio aerials, hydraulic cranes, camera lenses, lipstick tubes and other day-to-day objects. It is also invaluable in more specialized situations: photographers' unipods and tripods use it (in combination with hinging), as do the height-adjustable lights used on film sets and in television studios. And finally it has some rather curious applications, including telescopic plastic mugs and an expandable bullet-proof belt.

ACHROMATIC POCKET TELESCOPE

CHEAPEST 3 DRAWER BRASS TELESCOPE ON MARKET

AN IDEAL TELESCOPE **TAKES UP LITTLE ROOM**

Positively the finest **Pocket Telescope** on the market. It is made in France by a very large manufacturer of high grade Optical Appliances. It is fitted with **seven lenses** of high quality and power, rendering objects many miles away to be seen with clearness. When fully extended it measures almost fifteen inches long, and a little over five inches when closed. A feature which distinguishes this, as with all of the Brass Telescopes that we catalog, is its beautiful finish, the drawers and fittings being entirely of brass, exquisitely burnished and polished, with leather covered body, giving it an appearance of richness that shows it at once to be a superior glass. A brass cap at one end and special slide on the other exclude the dust when the instrument is not in use. Complete in neat cloth covered case.

No. 9111. Achromatic Pocket Telescope, postpaid **$4.95**

The mechanical principle of the telescope lives on, but advances in optical science have made a museum-piece of the telescope itself, such as this one advertised in the 1929 Johnson Smith & Co. catalogue.

Didactic telescoping: a lecturer's pointing device.

Film lights mounted on stands that slide up and down, telescope-fashion.

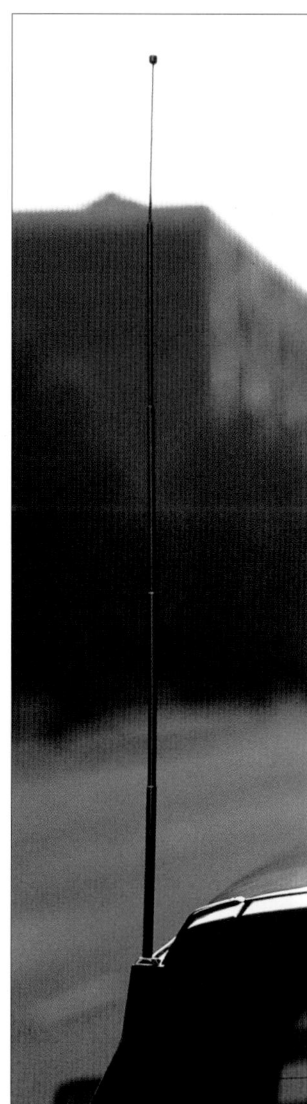

Telescope car aerial.

Double take: the Walkstool from Scandinavian Touch of Sweden extends to two heights, and folds down for easy carrying. Designed by M. Gustafsson.

Telescoping crane.

Telemax sliding ladder.

Enticer telescopic fishing rod by Abu Garcia of Sweden. It extends to a full 5 m (16 ft 5 in).

The Leica IIIG, 1938, by Leica, Germany.

The size of a camera is basically determined by the size of the film roll, the frame and the lens. In principle, the larger the frame and lens diameters, the sharper the picture. But a bigger lens requires a greater distance between the lens and the film. That leaves two ways to reduce the size of a camera. One way is to collapse the distance between the lens and film when the camera is not in use. Older cameras often had bellows to achieve this (see pp. 72–73). The other way is to improve the quality of film and lens. This is what the revolutionary Leica camera did. Designed by Oskar Barnack, it had a huge impact on the market and history of the camera. Essentially, Leica invented the small camera with small negatives for enlarged prints. The idea was to reduce camera size without reducing picture quality. Collapsibility was part of the solution. The first Leica camera, which went on the market in 1925, combined three innovations. First, it used a roll of 35 mm film to produce multiple 24 x 26 mm negatives. Second, it introduced the powerful 50 mm f:3.5 Elmar lens. Third, the lens was tube-mounted and slid back into the camera body when not in use.

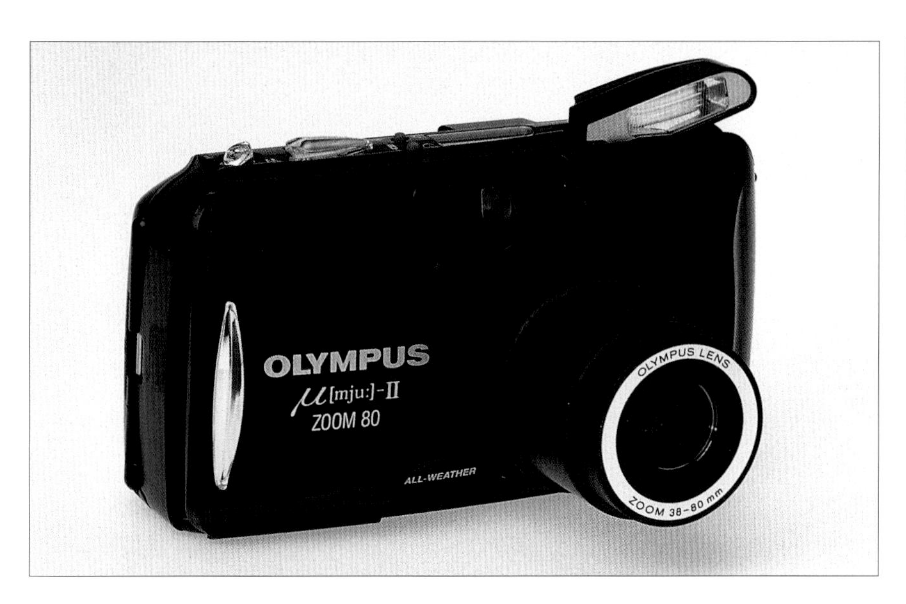

Like the Japanese Olympus model, many contemporary autofocus pocket cameras increase and reduce the distance between lens and film by the telescope principle.

Gentlemen's jewelry: the Minox 8 x 11 mm camera is a photographic notebook that can be taken everywhere – undercover if you happen to be a spy (or fancy yourself as one). The camera slides open to expose the viewfinder and lens, and slides back and forth to wind on the film. It is stored closed to protect the lens. Since its launch in 1935, the camera has undergone very few changes – though today, naturally, it is available in a matt black finish.

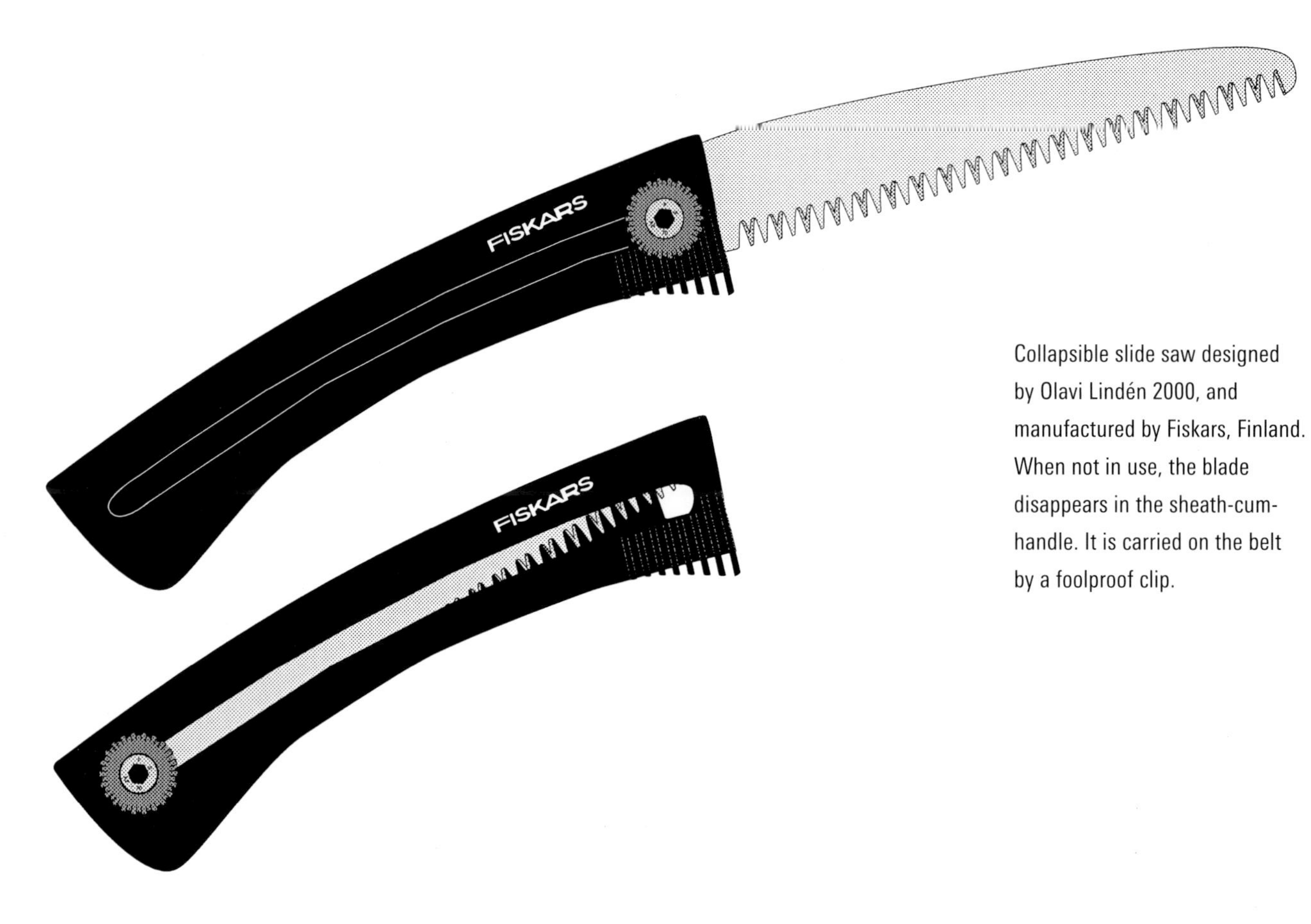

Collapsible slide saw designed by Olavi Lindén 2000, and manufactured by Fiskars, Finland. When not in use, the blade disappears in the sheath-cum-handle. It is carried on the belt by a foolproof clip.

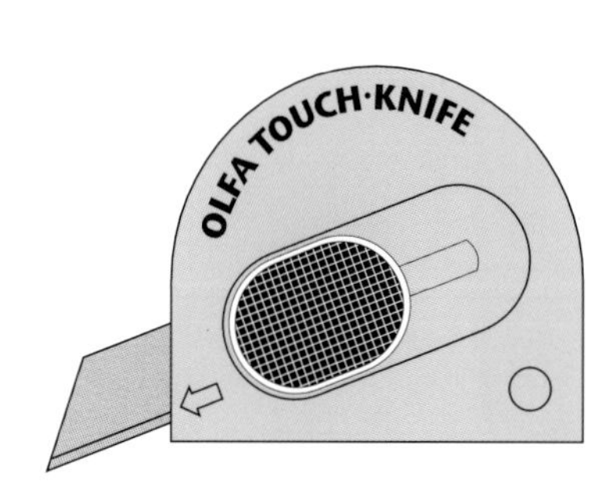

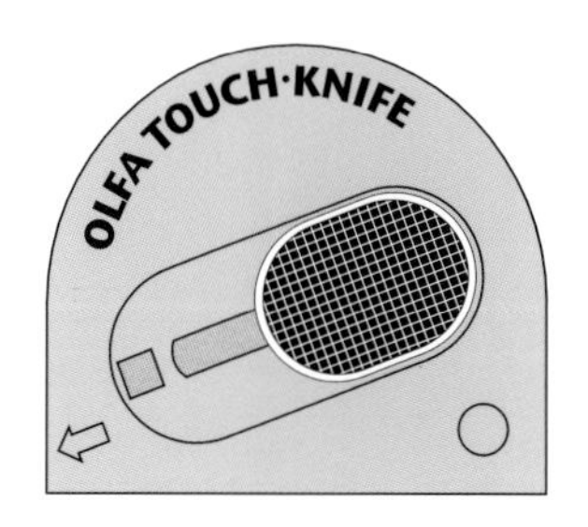

The Olfa Touch-Knife paper cutter, with a blade that slides out for use and back for safe storage.

Women – and occasionally men – have coloured their lips since the dawn of civilization. But it was not until 1915 that the lipstick as we know it today – a stick of coloured wax that slides in and out of a containing tube – was introduced to the US market. In 1770 the British parliament passed a law against the artificial colouring of lips, and declared that women found guilty of seducing men into marriage by such deceit could be tried for witchcraft. The bullet shape of an extended lipstick has many times been noted. Lipsticks are now big business, and many women in the West buy several bullets each year.

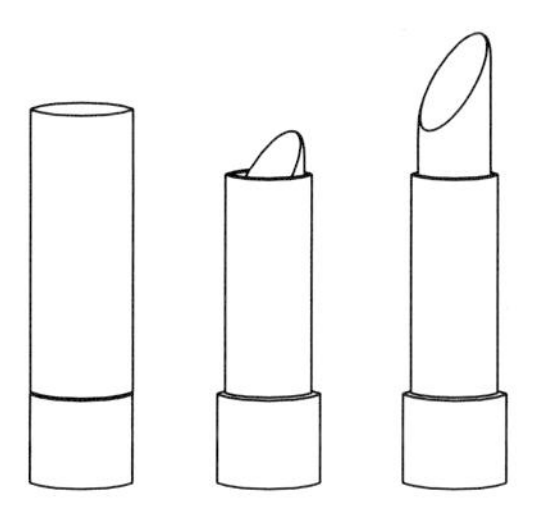

Always keep one eye on the professionals. Flight crews are not allowed to appear stressed. That's why they never carry personal luggage long distances through an airport. Instead they use small carts or wheeled luggage with pull-out telescopic handles, like this case from Tumi Luggage, US. Watch and learn.

Telescopic mugs, in luxury silver with leather cases (top and middle) and in plastic (below).

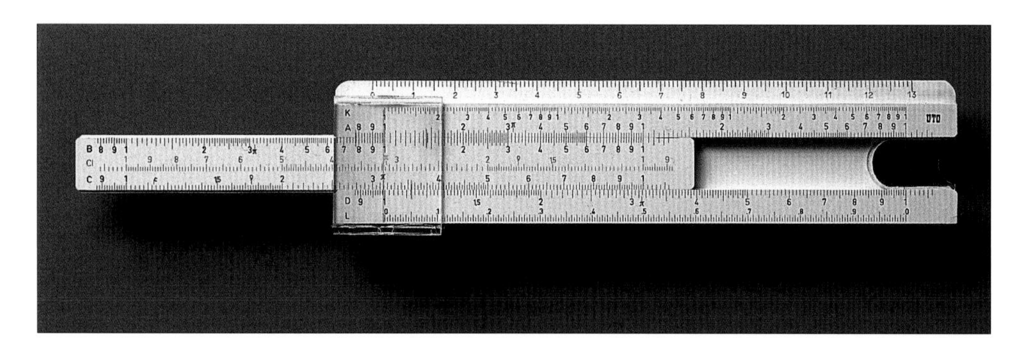

Video killed the radio star. And the slide rule didn't stand a chance against cheaper, smaller, faster, more versatile and more precise pocket calculators.

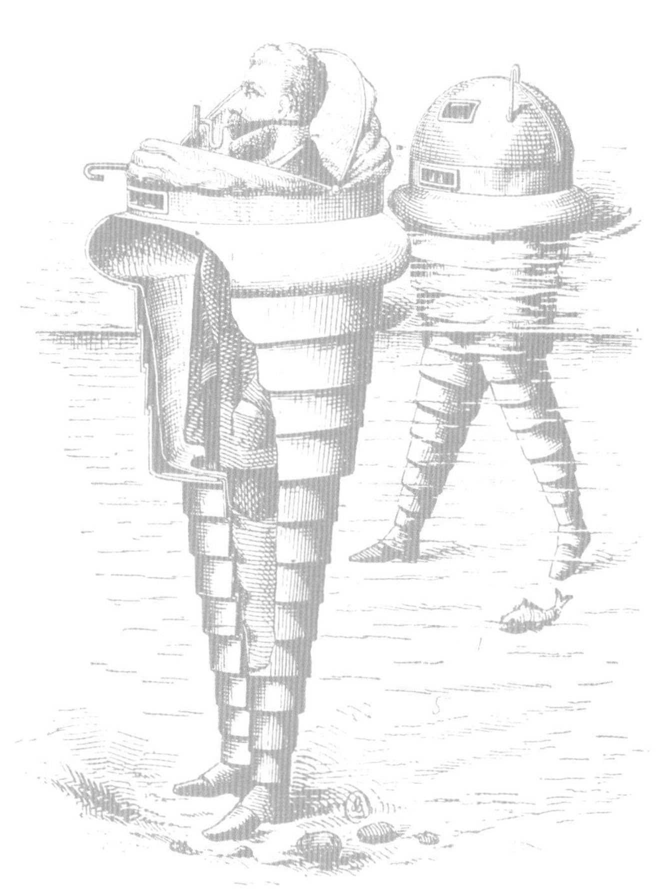

Japanese Chochin *kabuto* or helmet, Samurai armour from the Edo period (1603–1868), which telescopes almost flat for storage.

This life-preserver for ship-wreck victims was designed by Traugott Beek and appeared in *Scientific American* in 1877. The upper cavity holds a month's supply of food and drinking water.

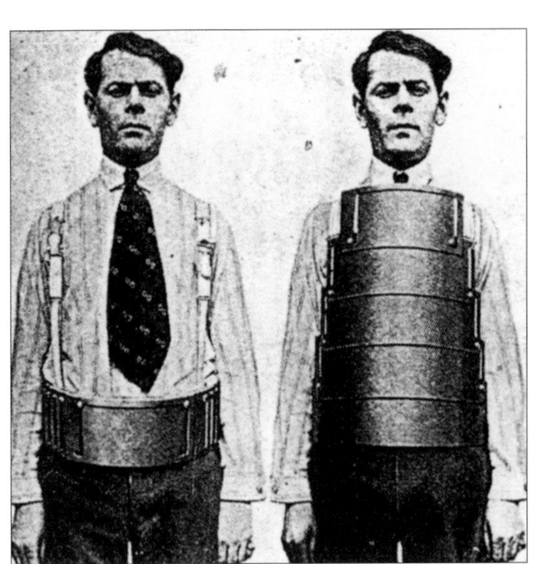

Telescopic protection against bullets that folds down into a 3.5-kilo (7 lb 12 oz) metal belt. Illustrated in *Popular Mechanics* in 1918.

Nesting

Russian *matryoshka* dolls that fit one inside the other are the perfect prototype of the nesting principle.

Nesting is a group principle. Together, two or more nested objects occupy less practical space than they do individually. If an object is to be capable of nesting, it must have some kind of cavity that can be occupied by another object. The result is negative space synergy: one plus one equals one and a half.

Spoons of equal or diminishing size will nest if they are carefully arranged. If left in a mess, on the other hand, they take up far more space and may hardly fit in the drawer. Measuring spoons often come in sets of diminishing size which fit neatly into one another. Many other kitchen tools nest too: pans, plates, cups and more.

Objects do not need to have an identical shape to nest. In the early 20th-century traveller's tea set on p. 149, burner, cup, sugar bowl and cream jug are stored inside the tea pot and/or each other.

Lots of everyday utilitarian items nest conveniently when not in use – orange traffic cones or supermarket trolleys and shopping baskets. Builders and demolition workers use nesting rubbish chutes to dispose safely of unwanted debris. Nests of tables nest, and many chairs are designed to stack, a variation on the nesting principle.

When appointed Cultural Capital of Europe in 1998, the city of Stockholm commissioned a number of talented young designers to create lines of quality souvenirs. Among those chosen for production were two nesting collapsibles: a set of cutlery designed by Glenn Roll (above) and a set of brushes by Pia Kristoffersson (opposite).

Good Grips nesting measuring spoons manufactured by OXO in the US.

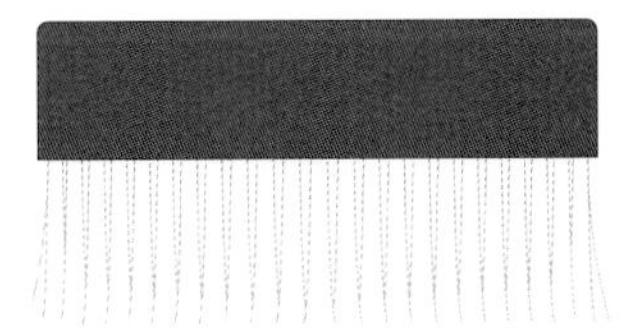

Set of three nesting brushes designed by Pia Kristoffersson to commemorate Stockholm's year as Cultural Capital of Europe in 1998.

Margrethe nesting kitchen bowls, designed in 1955 by Jacob Jensen for Acton Bjørn & Sigvard Bernadotte and manufactured by Rosti. In 1998 they were featured in a series of Danish stamps celebrating good design.

Klubi stacking glasses designed by Harri Koskinen in 1998 for iittala, Finland.

Aluminium dessert moulds.

An elegant solution, if only modestly space-saving. Melamine stacking egg cups designed by Kristian Vedel in 1960 for Torben Ørskov, Denmark.

Unpacking this nesting tea set manufactured around 1910 in Berlin is half the pleasure of a nice cup of tea en route. Making up the party are a pot with lid and handle, mug, strainer, cream jug, base and burner.

NETTO

◀

Empty shopping baskets piled high and begging to be filled. They help shoppers carry their goods before reaching the till, they protect the merchandise, and they aid the store detective's discreet surveillance.

The Dutch inventors of the building chute, G.H. and A.J. Vlutters, took their inspiration from stacks of Dixie Cups. (See the point of this book?) They patented their invention in 1984. Since then innumerable chutes have carried debris from scaffolds directly down into containers on the street without risking the safety of pedestrians or other road users.

Horizontally nested shopping trolleys.

I Love You

◀

Children love the helium-filled collapsibles sold by the balloon man in the Tivoli Gardens of Copenhagen. Parents fear his inflated prices.

US Air Force aerial, produced by Goodyear in 1960. An energetic soldier with a foot-pump could inflate it to its full height of 18 m (59 ft) in 15 minutes. Yes Sir!

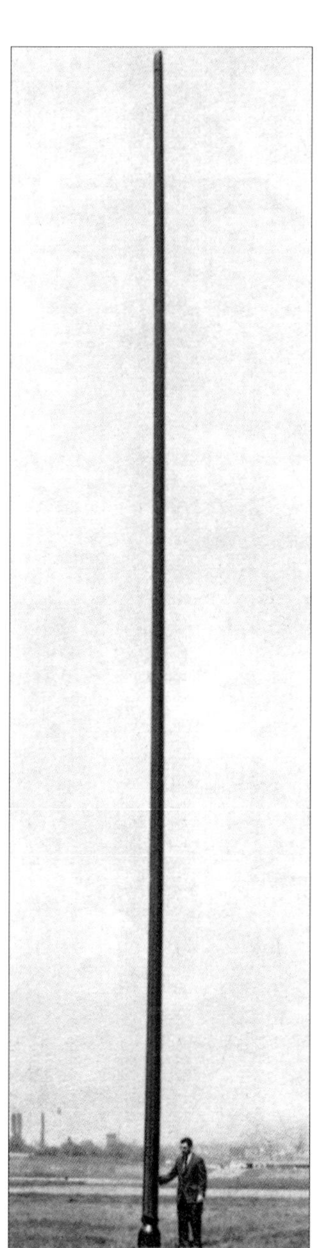

Inflatable collapsibles: hot-air balloons are the largest; children's balloons are the most common; life rafts and life jackets are the most useful; inflatable postcards are the least useful; inflatable furniture is the most problematic; an 18-metre (59-foot) radio aerial is the most incredible; a blow-up doll is surely the most bizarre; and children's inflatable beach toys are perhaps the most entertaining.

The concept of inflation is two-fold. On the one hand we associate it with a normally undesirable economic phenomenon. On the other it is a structural principle whose potential is limited only by the imagination. An inflated mini-hovercraft that is more off-road than most off-roaders conjures all kinds of cartoon fantasies.

Architecture can be inflatable too. Playground structures offer children an entirely novel experience: the interactive floor. Pneumatic buildings used to be simply tethered inflatables, but now there are tents supported by inflated arches. These eliminate the need for air-lock chambers at the entrance. Such inflated tents are used for both military and civilian purposes. Entire hospitals can be 'built' in an incredibly short time, which is invaluable in emergency situations.

OceanSpray®

▲
The Mongolfier brothers' hot-air balloon, in which the world's first untethered flight was made.

◄
Contemporary hot-air balloons in mid-air traffic congestion.

►
A hot-air advertising balloon made by Cameron Balloons in the UK is deflated to be taken away by a small truck.

A hot-air balloon is a lighter-than-air craft without a propulsion system. In France it is called a *Mongolfière* after its French inventors, brothers Joseph and Étienne Mongolfier. They made the first manned flight over Paris on 21 November 1783. Early balloons were filled with hot air from a fire on the ground before take-off, and descended when the air temperature fell.
On-board burners were a later development. They allow greater control over the temperature and so the altitude of the balloon. Other types of balloon have avoided the dangers of fire or burner by using gases lighter than air. French physicist J.A.C. Charles made a two-hour flight in a hydrogen-filled balloon just ten days after the Mongolfiers made their first ascent. Today, excursions by hot-air balloon are usually funded by advertising sponsorship. Light gas balloons are still used, as weather balloons, radio transmitters and as permanent advertisements.

OCEAN
ATLANTIC
OCEAN
SUNDAY, HAMBURG

Inflatable comfort: the traveller's neck cushion.

Inflatable convenience: blow-up washing-up bowl.

Pneumatic grooming: an inflatable coat hanger.

Air mail.

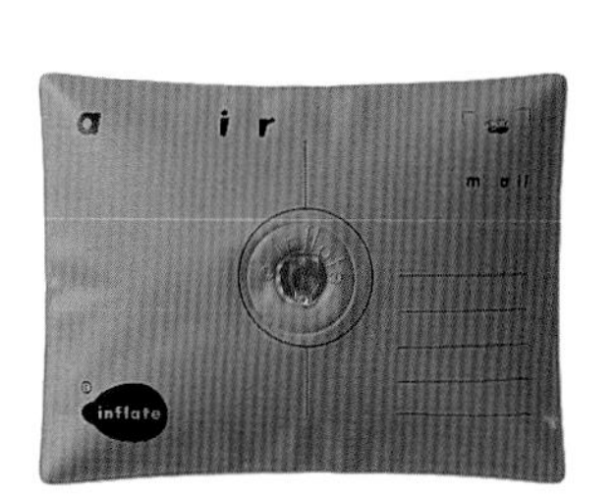

Buoyancy for your beer: an inflatable Carlsberg cooler.

◄

Ever since Charlie Chaplin filmed *The Great Dictator*, we can't help thinking of the earth as a balloon.

Inflatable life jackets, rafts and aircraft escape slides form a special category of survival design for emergency situations. They must be self-explanatory or – better still – self-activating in circumstances where not all users can be expected to be fit to fend for themselves.

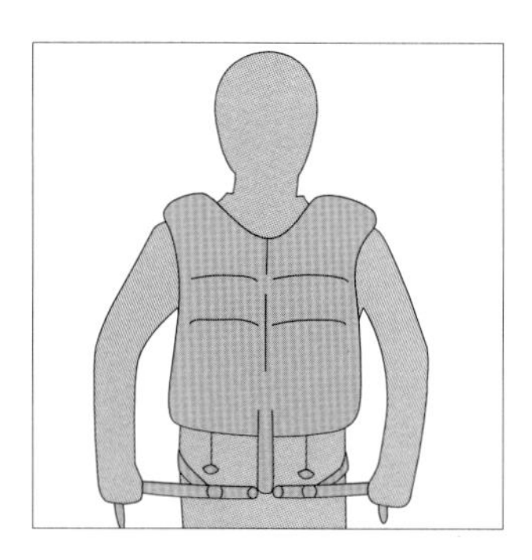

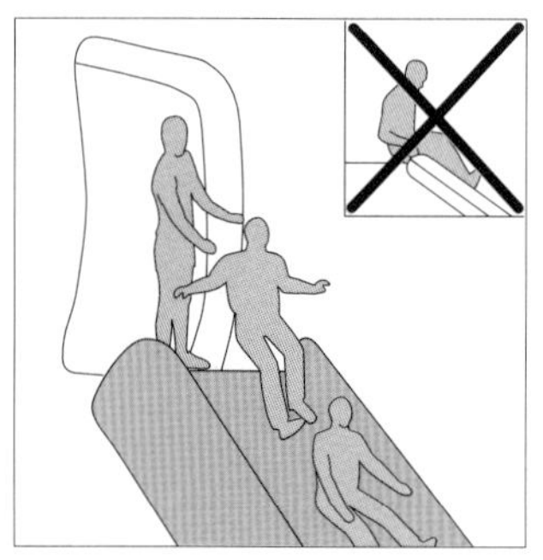

In the event of an emergency landing, aircraft passengers depend on an inflatable chute to slide and/or float to safety.

The Viking life raft, manufactured by Viking Life-Saving Equipment, Denmark.

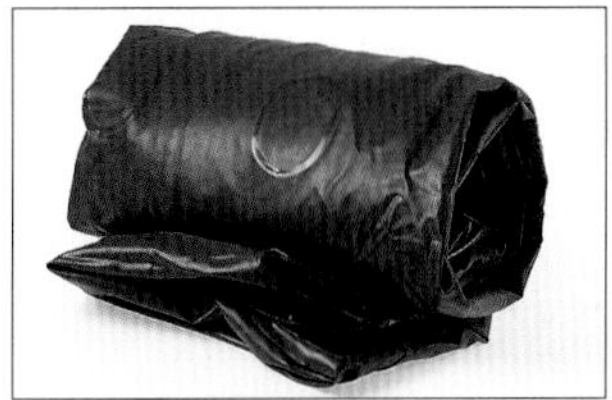

The motorized self-inflating bed sold by Quantum International in the UK will always beat straining over a foot pump when you are worn out.

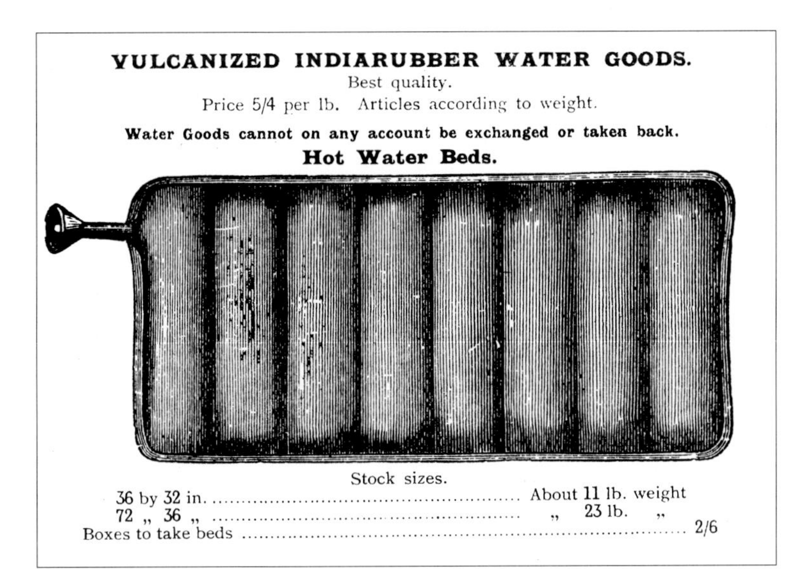

VULCANIZED INDIARUBBER WATER GOODS.

Best quality.

Price 5/4 per lb. Articles according to weight.

Water Goods cannot on any account be exchanged or taken back.

Hot Water Beds.

Stock sizes.

36 by 32 in. .... About 11 lb. weight

72 „ 36 „ .... „ 23 lb. „

Boxes to take beds 2/6

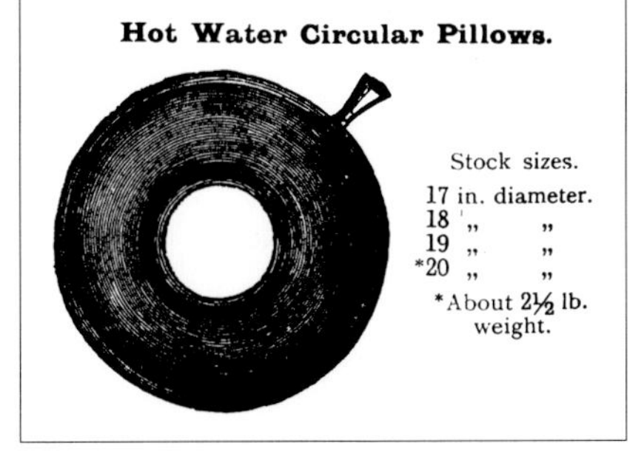

Hot Water Circular Pillows.

Stock sizes.

17 in. diameter.

18 „ „

19 „ „

*20 „ „

*About 2½ lb. weight.

Strictly speaking, inflatability refers to air or gas, but flexible water cushions work in much the same way. These hot-water inflatables were sold in the Army & Navy Stores catalogue in 1907.

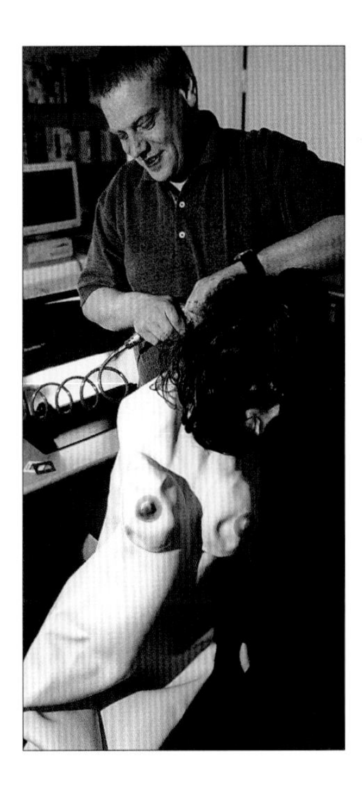

The sailor's mate: companion of last resort.

Float away on a tropical synthetic.

The future has arrived: this mini-hovercraft inflates from the exhaust of a 12-horsepower engine and cruises over land, grass, water, marsh, ice, mud, sand and snow – whatever you find. And it fits into the boot of a car. Hovery, designed by Alberto dei Castelli and manufactured by Argentinian Aero DC.

▲

Inflatable architecture: the Wet Set beach life. Manufactured by Intex, US.

◄

Inflatable flooring. Tivoli Gardens, Copenhagen.

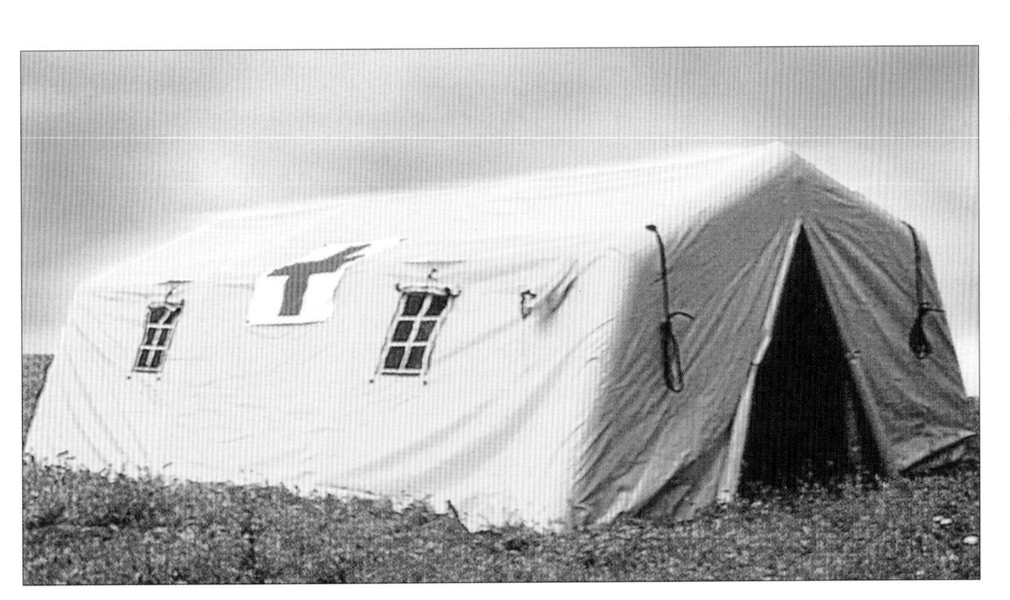

Instant architecture: tent with pneumatic structure that can be rapidly assembled by four people. Four arches are inflated by compressors or manual pumps. The tent, made of reinforced plastomer fabric and manufactured by Italian firm Eurovinil, was designed for civilian first aid in disaster zones.

Fanning

Fans were once an essential feminine accessory. These days you are more likely to see a so-called fan of playing cards. A true fan has a pivot that holds its leaves together and allows multiple leaves to be viewed at the same time.

A fan is a useful way of keeping together a bundle of flat – or flattish – complementary or comparable items, both for display and storage. Thus colour swatches often fan open on a pivot, and allen keys and other tools may come in a fan mounting.

Fans may also have less utilitarian uses. The curtain of the Pantomime Theatre of Tivoli, the amusement park in downtown Copenhagen, is a peacock that shows off its splendid tail at curtain-fall.

The Herd Tametoma Sails Home from the Ryukyu Islands, woodcut by Kuniyoshi (1798–1861).

Fans have sometimes been more than temperature-controlling devices; they can be an elegant aid to coy flirtation. These, sold in the Army & Navy Stores catalogue of 1907, were made of feathers, lace, bone and pearl.

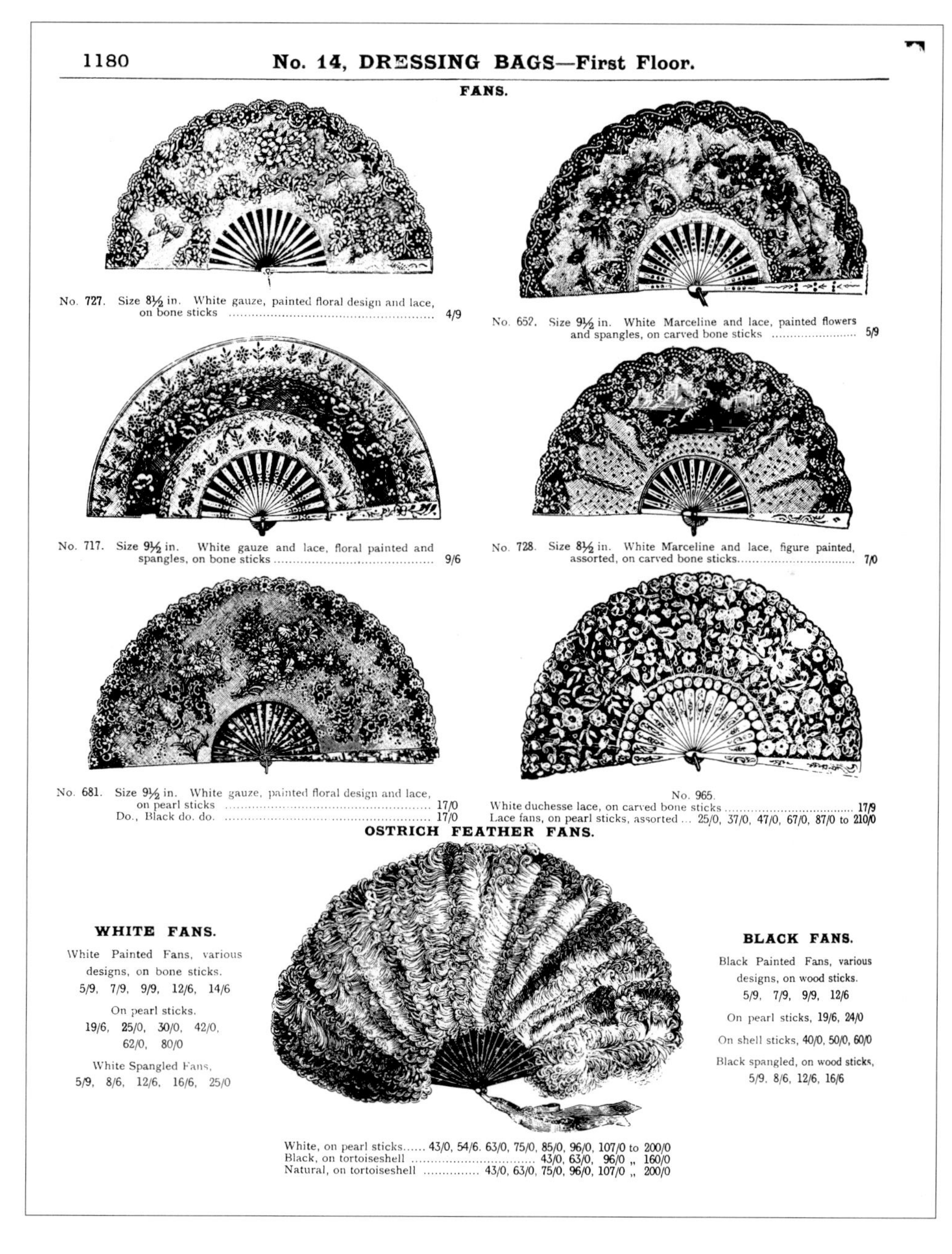

1180 **No. 14, DRESSING BAGS—First Floor.**

FANS.

No. 727. Size 8½ in. White gauze, painted floral design and lace, on bone sticks 4/9

No. 652. Size 9½ in. White Marceline and lace, painted flowers and spangles, on carved bone sticks 5/9

No. 717. Size 9½ in. White gauze and lace, floral painted and spangles, on bone sticks 9/6

No. 728. Size 8½ in. White Marceline and lace, figure painted, assorted, on carved bone sticks 7/0

No. 681. Size 9½ in. White gauze, painted floral design and lace, on pearl sticks 17/0
Do., Black do. do. 17/0

No. 965.
White duchesse lace, on carved bone sticks 17/9
Lace fans, on pearl sticks, assorted ... 25/0, 37/0, 47/0, 67/0, 87/0 to 210/0

OSTRICH FEATHER FANS.

WHITE FANS.

White Painted Fans, various designs, on bone sticks.
5/9, 7/9, 9/9, 12/6, 14/6

On pearl sticks.
19/6, 25/0, 30/0, 42/0, 62/0, 80/0

White Spangled Fans,
5/9, 8/6, 12/6, 16/6, 25/0

BLACK FANS.

Black Painted Fans, various designs, on wood sticks.
5/9, 7/9, 9/9, 12/6

On pearl sticks, 19/6, 24/0

On shell sticks, 40/0, 50/0, 60/0

Black spangled, on wood sticks,
5/9, 8/6, 12/6, 16/6

White, on pearl sticks...... 43/0, 54/6, 63/0, 75/0, 85/0, 96/0, 107/0 to 200/0
Black, on tortoiseshell .... 43/0, 63/0, 96/0 „ 160/0
Natural, on tortoiseshell .... 43/0, 63/0, 75/0, 96/0, 107/0 „ 200/0

Coolness and shade combined, in a fan-cum-hat from Manila.

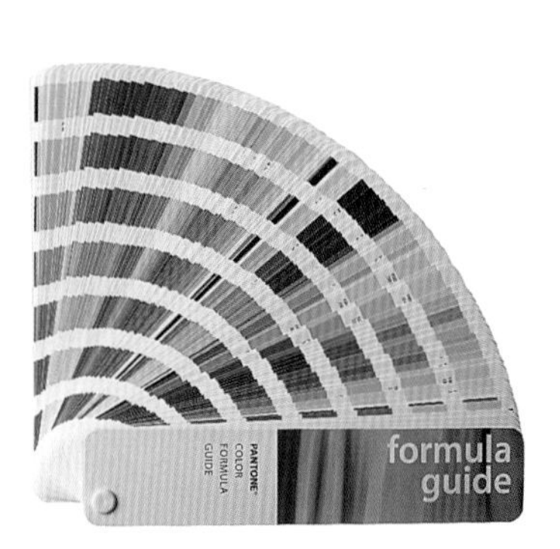

Pivoted colour swatches from Pantone and 3M.

Fan-mounted allen keys.

The spectacular curtain of the Pantomime Theatre of the Tivoli Gardens in Copenhagen is a giant fan in the form of a peacock. The building was designed by Vilhelm Dahlerup in 1874. The idea for the peacock curtain was Bernhard Olsen's, Tivoli director, after seeing a fan curtain in Paris.

Now, as then, the curtain is manually operated. Five strong men exercise their scenic powers in the cellars beneath the theatre.

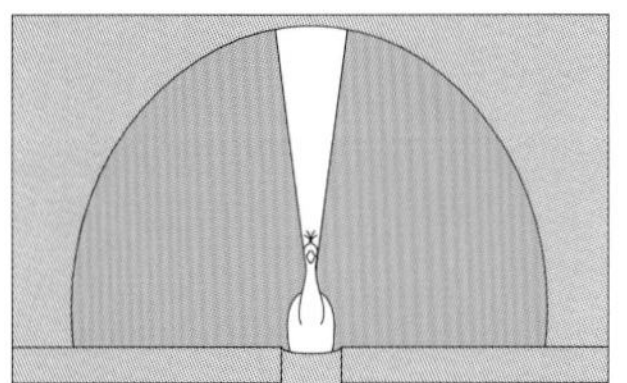

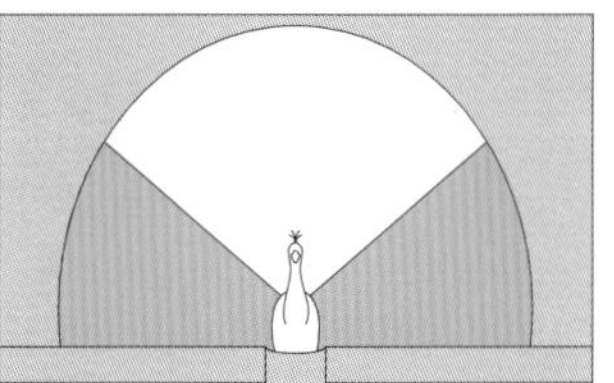

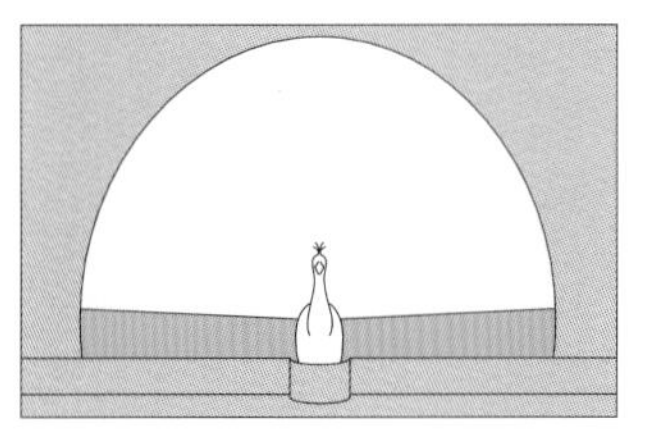

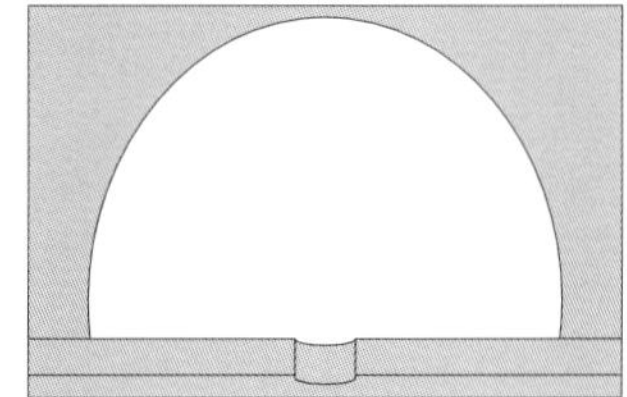

TC7
24
TC7
22
BBC 2

Concertina

◄

Concertina-mounted spot lights in the BBC television studios.

The concertina principle is a misnomer: the musical instrument to which it bears superficial similarity and from which it derives its name is in fact an application of the bellows principle.

Concertina collapsibles have a number of equal rods connected by pivots to form a string of Xs – XXXXX – which can be expanded and retracted by changing the angles between them.

The concertina principle allows wall-mounted adjustable mirrors and lamps to be manoeuvred. In TV studios, technicians pull down concertina-mounted monitors and lights from the ceilings. In some offices computers are drawn up into silos in the ceiling for extra data security outside office hours. The same system also allows the height of work stations to be adjusted to the needs of each user.

In the past, shopkeepers would often have a pair of concertina tongs to reach for objects in the far corner of the window display. Such tongs are still invaluable to people of impaired mobility.

Morgues and emergency wards use height-adjustable stainless steel tables with concertina bases, sometimes these days recycled for the fashionable living room. Italian housewives living over narrow alleys hang out their laundry to dry on concertina racks. And the principle is found in all kinds of other daily applications, from corkscrews to floppy disk holders.

Concertina-mounted retractable wall lamp by Le Klint, Denmark.

Expandable candle stick from the Army & Navy Stores catalogue of 1907.

Venetian bed linen aired on a concertina rack.

Shop awnings often extend and retract on concertina-pivoted rods.

Lazy tongs for the elderly, disabled, and anyone who needs to reach beyond their limit.

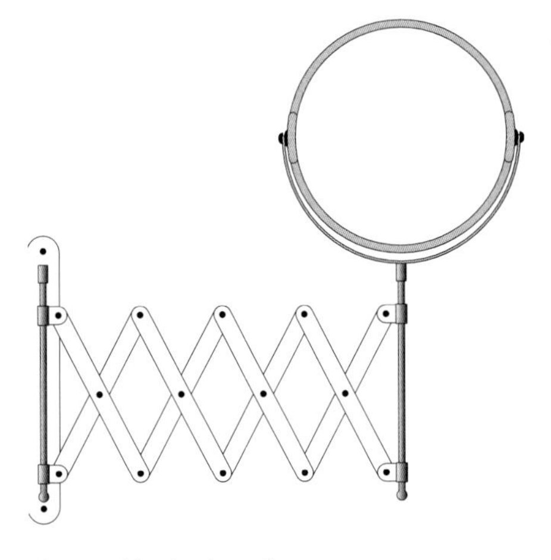

Retractable shaving mirror.

COLLAPSIBLE GARMENT HANGER

A big space saver for crowded wardrobes and closets in homes, hotels and business places. It holds the garments away from the wall, and yet it occupies no more space than one hook. It holds four times as much as any ordinary hook. Swings either way to give room. It can be put up on any wooden part, with screws, taking but a few moments. When not in use it can be closed up and thus it takes up hardly any room. It is a great space saver and it relieves crowded places. It is made of the best steel, is very strong and will hold up as many clothes as you can put on it. As there is nothing to wear out it will last a lifetime.

No. 4327. Collapsible Garment Hange 75c
3 for $2.00, or $6.75 per dozen Postpaid.

Concertina coat hanger, from the 1929 Johnson Smith & Co. catalogue, US.

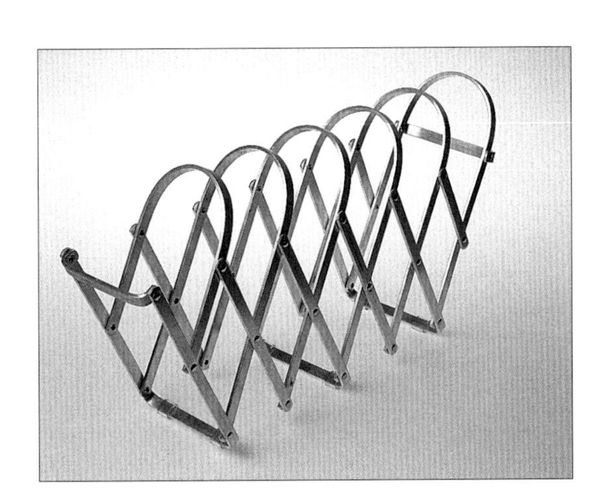

Length-adjustable paper sorter.

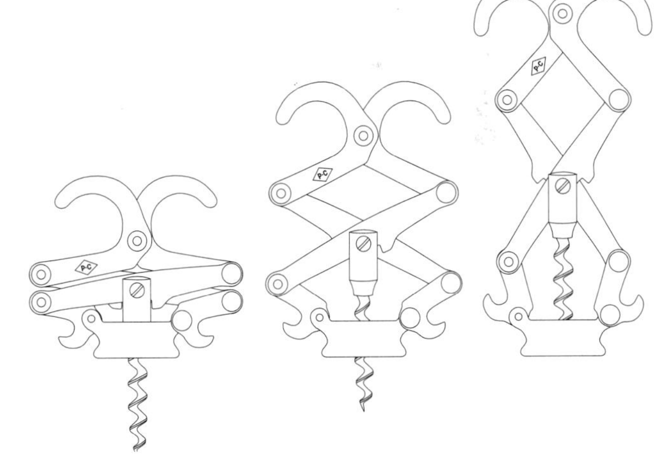

With a concertina corkscrew, the oenological pleasure starts while opening the bottle.

Retractable wall-mounted lamp originally manufactured in 1930s Vienna by Koranda.

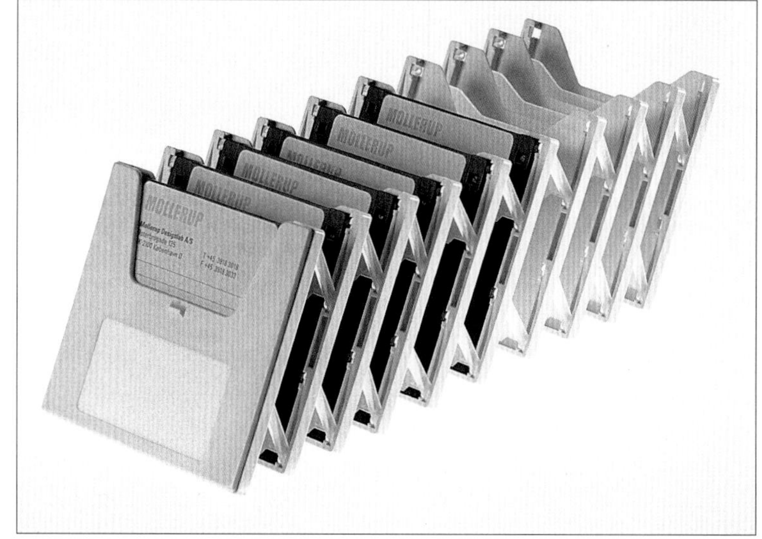

Concertina-like floppy disk cabinet that lets you scan through all your disks at once.

The Columbus foldaway staircase by Trip-Trap, Denmark.

Industrial elevator sold by Niels Bo, Denmark.

3. Furniture

Mobile functionality

Collapsible furniture

- Stools
- Chairs
- Beds
- Sofabeds
- Sofas
- Tables
- Storage furniture

Tools extend man's natural capacities. We hit harder with a hammer, drink more easily from a cup, and – most of the time – get along quicker by car.

A piece of furniture is a tool. By elevation and separation, it extends the capacities of the human body. Stools, chairs, beds and sofas are extra legs that raise man's body and offer him repose above ground level. Tables and case furniture are extra hands that elevate man's possessions over the ground. Case furniture also offers order and protection by separating its contents from the outside world.

The demand for collapsibility in furniture is created by either an intermittent need or a changing location of need. The Latin name for furniture, *mobilia*, expresses its mobile character, which may be enhanced by collapsibility. Much collapsible furniture has been designed for people on the move. Campaigning warlords have commissioned campaign furniture; safari hunters have commissioned safari furniture.

Campaign and safari furniture is often still produced, though for aesthetic rather than for practical reasons. The special function has become a sleeping function, outlived by its aesthetic expression: appearance over functional needs.

Collapsible carrying chair. Army & Navy Stores catalogue, 1907, UK.

Stools

Collapsible seating has been around for several thousand years – first as a symbol of authority, later as a practical tool. Egyptian dignitaries used collapsible stools in the second millennium BC. A folding stool, the sella curulis*, was both the seat and the official symbol of the Roman high magistrate. Later, the* faldistorium*, the throne of a Roman Catholic bishop, was collapsible too. These days, designing a collapsible stool is part of the apprenticeship of any ambitious young furniture designer, though this appears to be more a test of ingenuity than a response to market demand. A collapsible stool is a design challenge. Some of the most beautiful – if not necessarily the most useful – examples of modern Danish furniture are collapsible stools. Their designers include old furniture hands Kaare Klint, Ole Wanscher, Mogens Koch, Poul Kjærholm and Jørgen Gammelgaard.*

The X-frame stool is one of the oldest of all furniture types. This was found in Tutankhamun's tomb, Egypt, 1400–1350 BC.

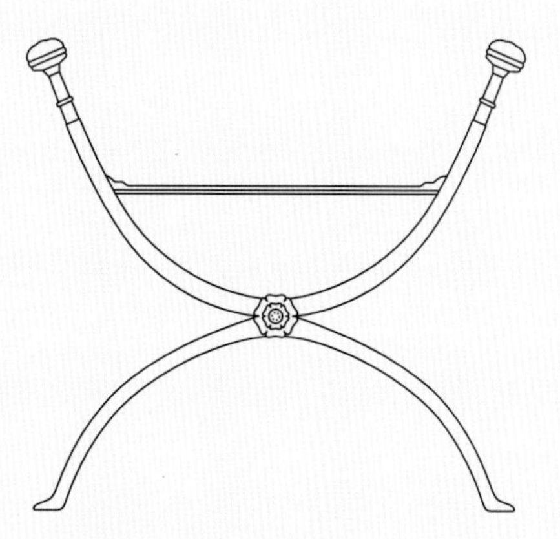

Recycling ancient symbols of power: a bishop's iron *faldistorium* from the cathedral of Bayeux, France, 14th century.

Roman coin issued in 40 BC in commemoration of Julius Caesar. The Roman Empire inherited the *sella curulis*, the folding seat of honour, from the Republic. It was a symbol of power, whose use was restricted to Roman emperors and the most important magistrates. The *sella curulis* was pictured on Roman coins alongside other symbols of dignity such as laurel wreaths, fasces and sceptres. One disgraced Roman praetor had his *sella curulis* broken into pieces. Caesar was murdered on his *sella aurea*, his golden stool.

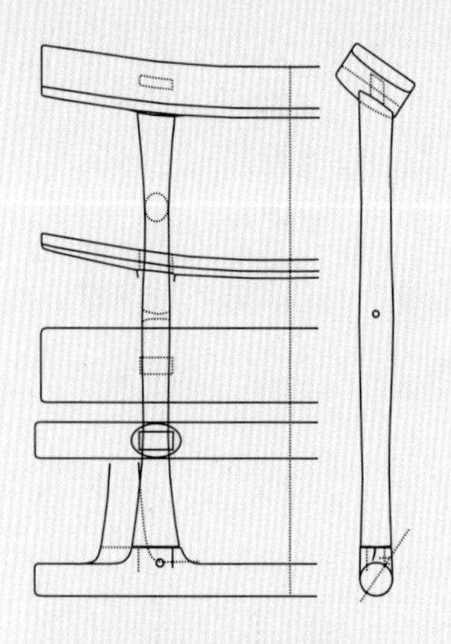

Egyptian folding stool, 1300 BC, measured and remade by Ole Wanscher in 1957.

Folding stool with propeller legs designed by Kaare Klint in 1930. The theme of twisted propeller legs was later taken up by other Danish furniture designers. Manufactured by Rud. Rasmussens Snedkerier, Denmark.

Folding stool *PK41* with propeller legs by Poul Kjærholm, 1961. Manufactured by Fritz Hansen, Denmark.

Folding stool with propeller legs by Jørgen Gammelgaard, 1970. Manufactured by Schiang, Denmark.

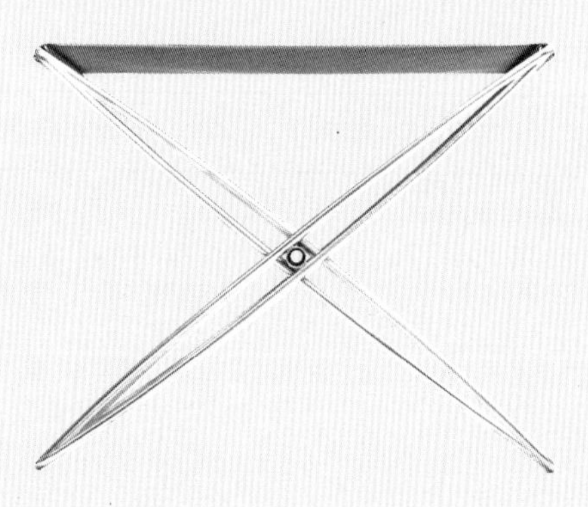

Folding stool by Mogens Koch, 1960. Manufactured by Rud. Rasmussens Snedkerier, Denmark.

Generic stool of Mexican origin. Manufactured by Form & Farve, Denmark.

The sky is the limit – almost – when Alvar Aalto's stools of laminated birch are stacked. Originally designed in 1933 with a seat of birch and three legs, the stool is also available today with vinyl-covered seat and four legs. Manufactured by Artek, Finland.

Stockholm by A&E Design (Hans Ehrich and Tom Ahlström), 1994. Manufactured by Zanotta, Italy.

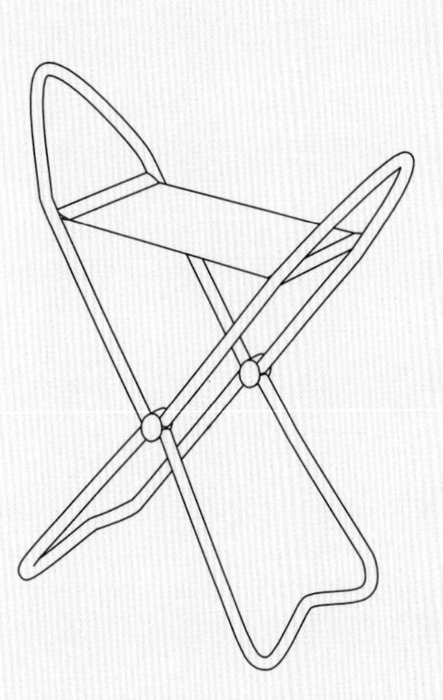

Chairs

Collapsible furniture includes a number of generic chair types that are available in both mass-produced low-cost versions and upmarket versions at ten times the price or more. The director's chair, the butterfly chair, the safari chair and the deck chair are examples of folk designs that have been taken up for upgrading by contemporary designers in pursuit of acclaim. After all, as Confucius said, 'to search for the old is to find the new.' Most collapsible chairs fold flat for storage. Others, like Kaare Klint's safari chair, are assembled and dismantled. This requires rather more effort, so they tend to remain assembled; their collapsible capacity is generally dormant. The collective collapsibility of stacking chairs is an important type of nesting.

Alfred Hitchcock on location on his high director's chair.

The director's chair owes its name to its popularity among film directors. On set, the director's chair is a strictly personal status symbol, protected and individualized by the name written conspicuously across its backrest. Lesser movie makers sit on chairs bearing the title of the production. The same chair also goes by other names with roots in other usages: the captain's chair or the yacht chair. Whatever you call it, this type of collapsible chair has a scissors frame and folds side to side. It has been traced back to the late 18th century, and was probably used as campaign furniture in the American Civil War. Director's chairs are available in innumerable low-cost versions and in some carefully designed and manufactured upmarket models with changeable seats and backrests.

Director's chairs, normal and high. Manufactured by the Telescope Company, USA.

Designers take their inspiration wherever they find it. When Danish architect Mogens Koch moved into a new terraced house in Copenhagen in the 1920s, he bought a couple of cheap folding chairs with scissors frames in a nearby shop. When in 1932 he submitted designs to a competition for extra church chairs, he took up the scissors theme and improved on it. The advantage of Koch's design is that it won't trap the folder's fingers, unlike its source of inspiration. But the judges failed in their judgment, as they so often do: Koch's chair did not win the competition.

Mogens Koch's chair was later supplemented by a stool (see p. 177) and a children's chair. The latter was dubbed the *Grandchildren Chair* because only grandparents could afford it. The collapsible rack holds six collapsed chairs. Manufactured by Rud. Rasmussens Snedkerier, Denmark.

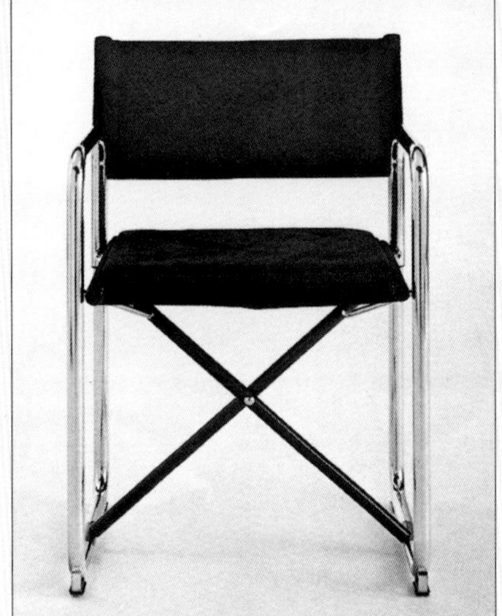

Åke Axelson's collapsible chair with wooden frame and armrests comes full circle: it is closer to Koch's buy than Koch's design.

The scissors-type folding chair with wooden frame cries out for a metal version. One answer is the *X-75* by Swedish designers Lindau & Lindecrantz. Manufactured by Lammhults, Sweden.

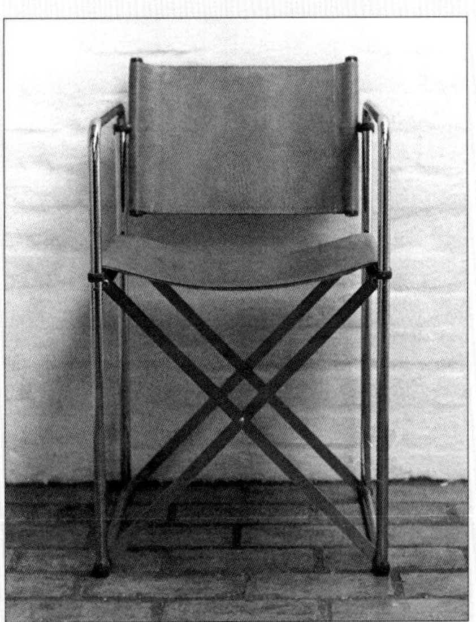

◂◂

While Lindau & Lindecrantz emphasized the strength of their material by only having one cross in the frame, Gae Aulenti went for double crossing with *April* in 1964. Manufactured by Zanotta, Italy.

Scissors-frame chair designed by Hans Bølling. Manufactured by Torben Ørskov, Denmark.

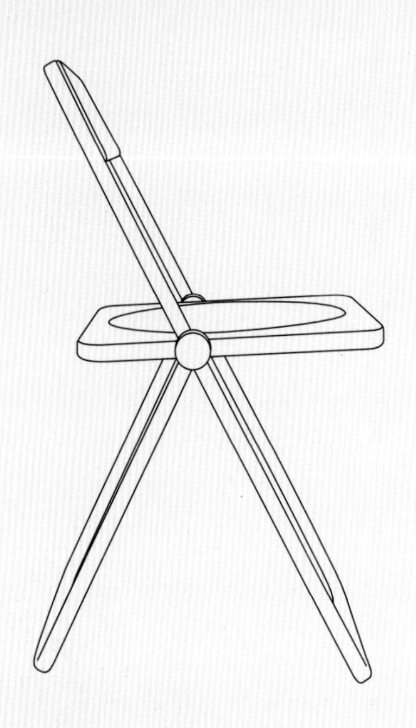

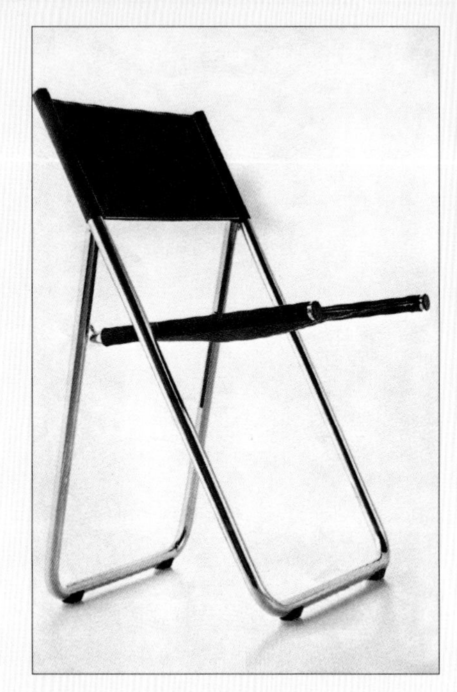

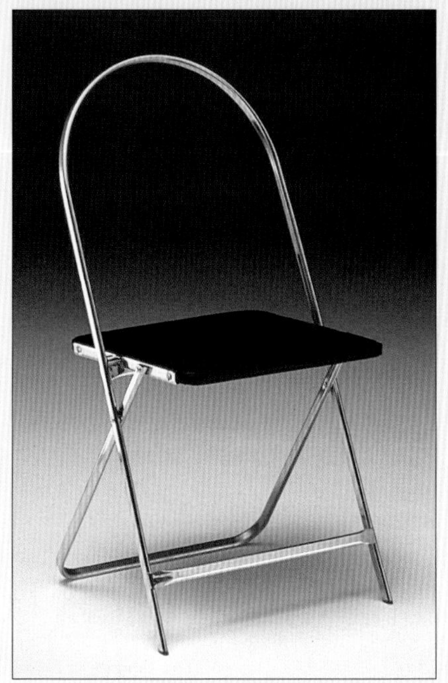

Since the whole idea of folding objects is size reduction, the concept of collapsibility inevitably implies a measure of minimalism. The *Plia* chair of chromed steel and moulded perspex designed by Giancarlo Piretti in 1969 is minimalistic both folded and unfolded. The see-through seat and back maximize the visual sense of minimalism. Manufactured by Anonima Castelli, Italy.

Lindau & Lindecrantz's idea of a minimalistic chair, *X-75-1*. Manufactured by Lammhults, Sweden.

Upgrading after reduction. *Aprilina*, in steel and leather, designed by Gae Aulenti in 1988, is a minimalist chair of similar design. Aprilina allegedly offers more comfort. Manufactured by Zanotta, Italy.

Nobody knows the exact origin of this generic outdoor chair, which is over a hundred years old.

In 1978 Marco Zanussi took up the anonymous design of the familiar outdoor chair for upmarket upgrading. He called the result *Celestina*. Yesteryear's seat and back of painted wood became a nylon seat and back covered with cowhide. Celestina is available with and without armrests. Manufactured by Zanotta, Italy.

Designing a collapsible chair in which seat, back and legs form a single X is a challenge taken up by many furniture designers. Hans J. Wegner's oak chair from 1949 folds beautifully, but probably seldom. For however slim and elegant it may be when adorning a room by hanging on the wall, most owners want to enjoy their well-designed possession in full. The tactile handles are in the same class as the knobs of a Shaker rocker. Manufactured by PP Møbler, Denmark.

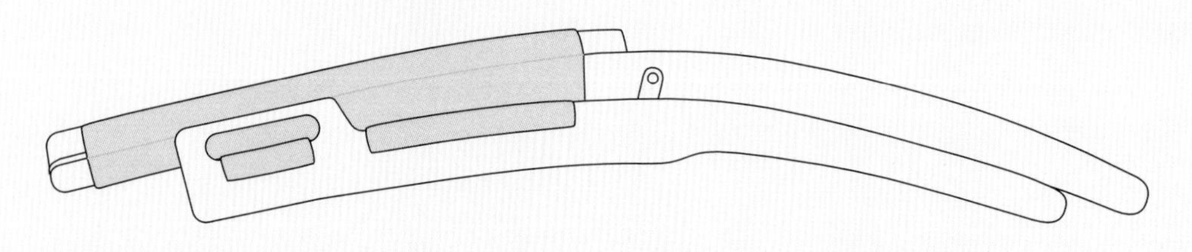

As so often before, Lindau & Lindecrantz did in steel what others did in wood. Their *X75-4* is adjustable and features a head cushion and newspaper pocket. Manufactured by Lammhults, Sweden.

Non-adjustable, no head cushion and no pocket, but absolutely minimalistic. Designer unknown.

Creativity is nothing but combining existing knowledge in new and useful ways. One brilliant X-chair solution is Ole Gjerløv Knudsen's beech and canvas *Bucksaw* chair of 1966, so-named because it is held together by the same principle as a bucksaw. Manufactured by Trip-Trap, Denmark.

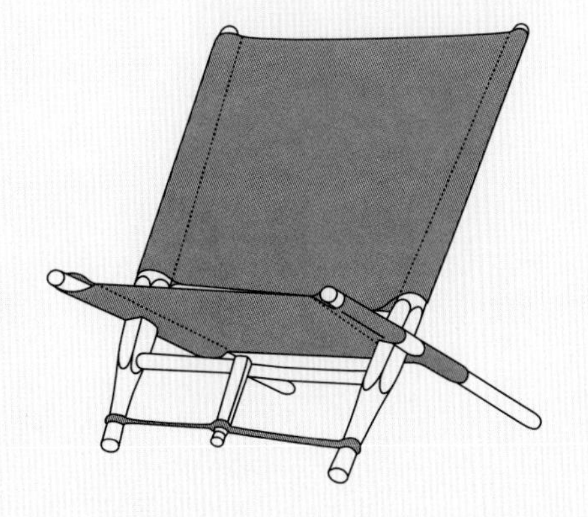

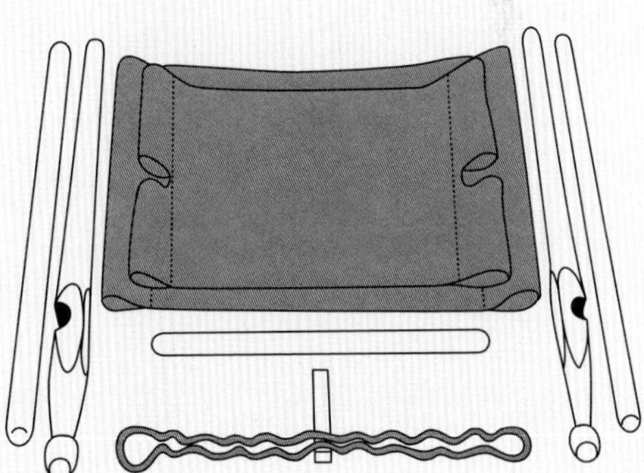

Furniture Chairs

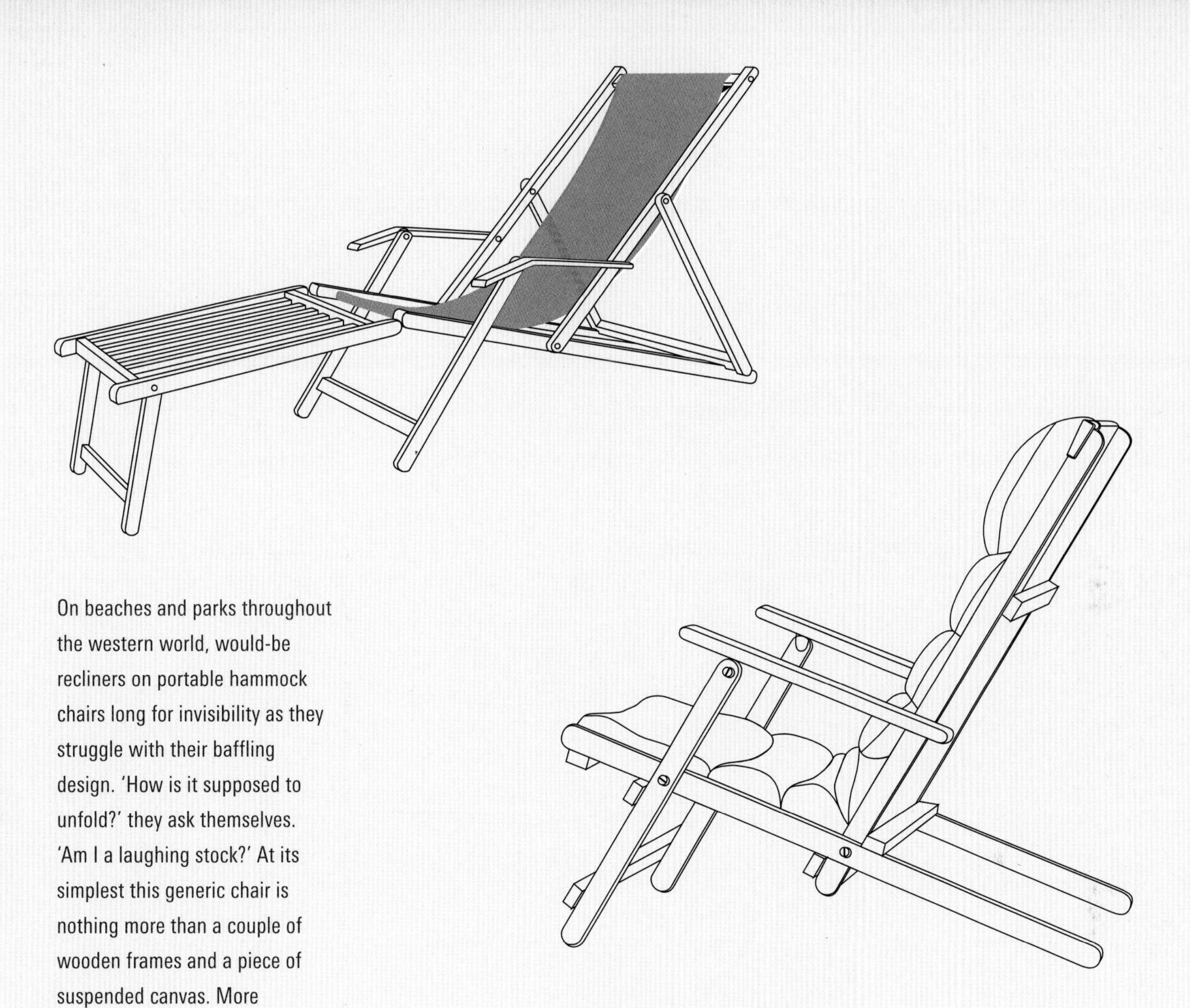

On beaches and parks throughout the western world, would-be recliners on portable hammock chairs long for invisibility as they struggle with their baffling design. 'How is it supposed to unfold?' they ask themselves. 'Am I a laughing stock?' At its simplest this generic chair is nothing more than a couple of wooden frames and a piece of suspended canvas. More elaborate versions may have armrests and footrest, and the back may be divided in two parts. Nobody knows how long it has been around or how many varieties exist.

Børge Mogensen's 1969 design is a substantial improvement on the simple hammock chair. Seat and back are separate, and the frame of the wooden seat is continuous with the back legs. The beechwood frame is connected by rivets that are galvanized to withstand the elements. Manufactured by Fredericia Furniture, Denmark.

Deck chairs are supposed to encourage relaxation and ease. However, they irritated Kaare Klint during a sea voyage by not folding in an aesthetically pleasing way. Back on shore, Klint designed his own deck chair that overcame the problem of untidy storage. Sadly the price of his perfectly folding deckchair makes it an eternal landlubber. Manufactured by Rud. Rasmussens Snedkerier, Denmark.

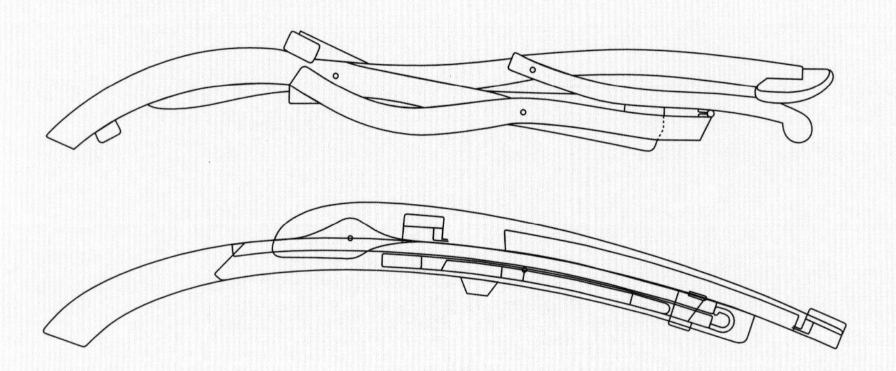

Folded conventional deck chair compared with Kaare Klint's elegant solution.

We still call it the safari chair, but it was never only used by safari-goers. It is a generic type of camp furniture dating back to the 19th century, and can also be known as a colonial chair or an officer's chair. In terms of collapsibility, it lags behind scissor-legged chairs. It doesn't fold as such; it dismantles and wraps up in itself. But that takes a while and you would only do it when moving from one base camp to another. Kaare Klint designed his safari chair in 1933, allegedly inspired by the photo of Martin and Osa Johnson on p. 15, which he saw in a travel book. Manufactured by Rud. Rasmussens Snedkerier, Denmark.

If Kaare Klint had looked in the 1907 Army & Navy Stores catalogue he might have noticed the *Roorkhee* chair.

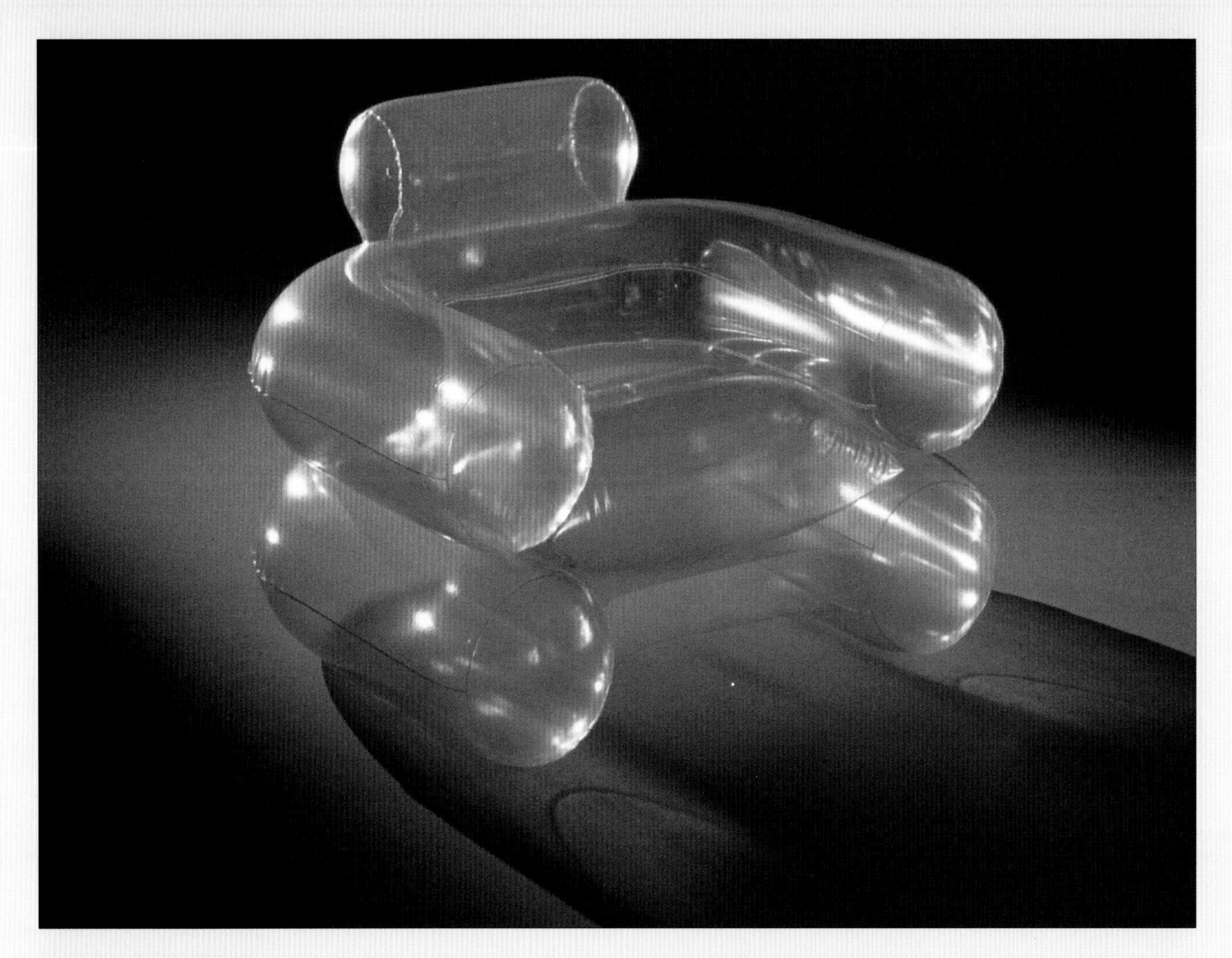

Blow was designed by De Pas, D'Urbino, Lomazzi and Scolari in 1967 as much as a visual joke as a functioning collapsible. It comes complete with air pump and repair kit. In form, Blow makes reference to Eileen Gray's *Bibendum* from 1929, which itself referred to Michelin's popular mascot. Manufactured by Zanotta, Italy.

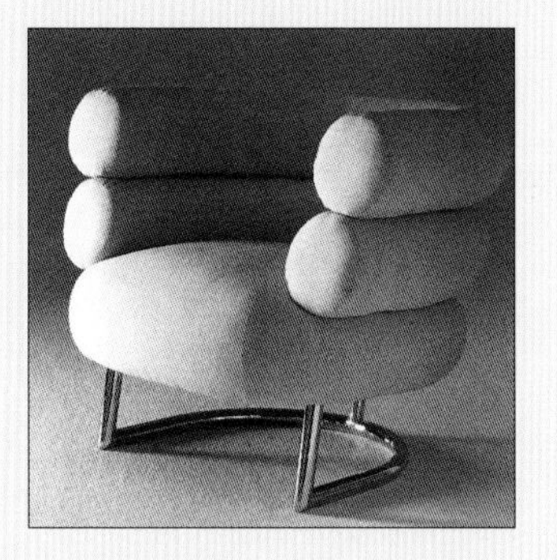

Blow's solid inspiration: *Bibendum* by Eileen Gray, 1929. Manufactured by Images, Italy.

IKEA don't usually waste printer's ink on telling customers who designs their furniture, but now and again some information escapes. The *a.i.r.* series of inflatable furniture was designed by Jan Dranger. It was a further development of designs made by Johan Huldt and Dranger for KF, the Swedish coop, in 1972.

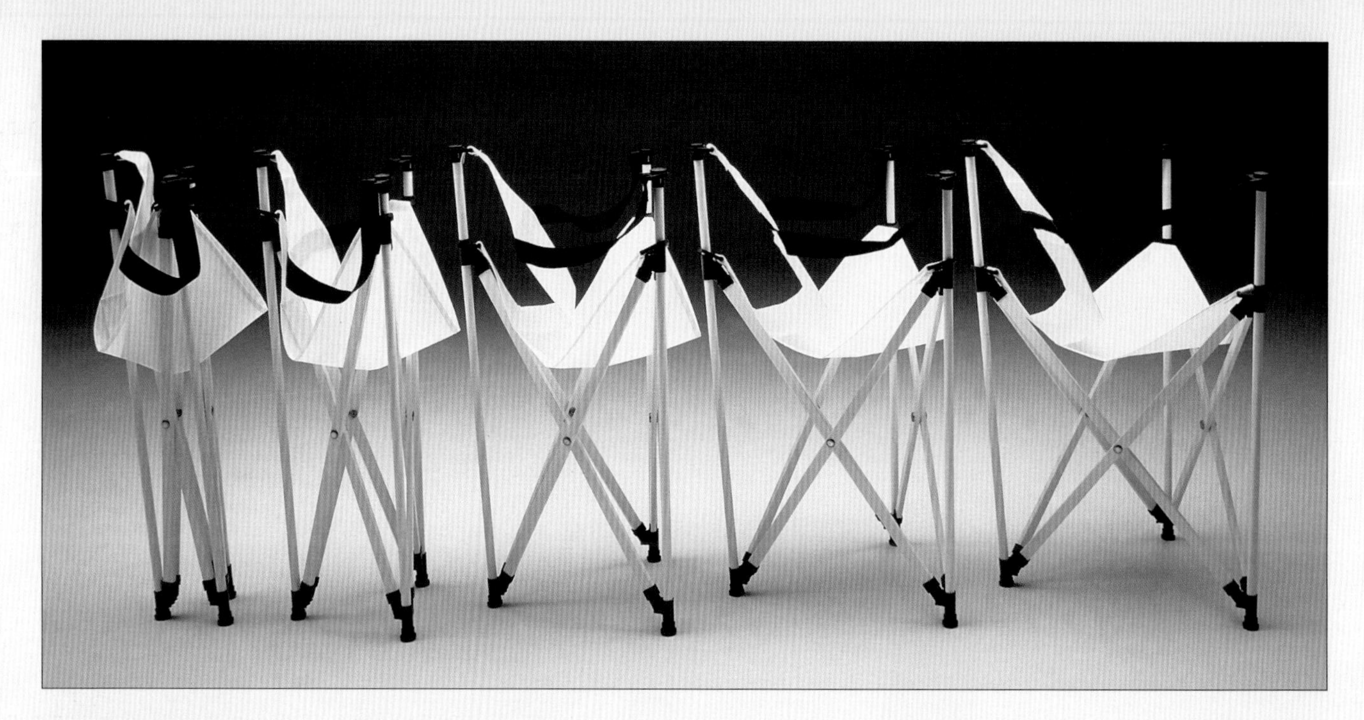

The *Trice* chair designed by Hannu Kähonen in 1986 exemplifies a group of furniture that is enjoyed more for its brilliant construction than its utility. Trice comes with its own (collapsible) bag for storage and transport. Manufactured by Amer Trading, Finland.

▶

Stocksessel, a walking stick-cum-seat from 1904. Manufactured by Thonet, Austria.

▶▶

Umbrella Chair by Gaetano Pesce, 1995. Another example of brilliant ingenuity over utility. Manufactured by Zerodisegno, Italy.

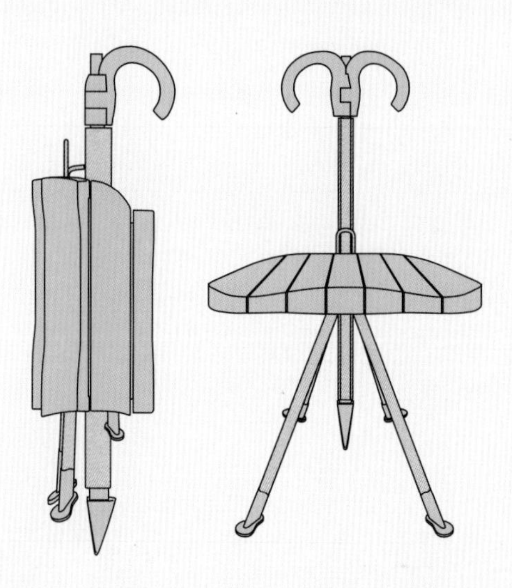

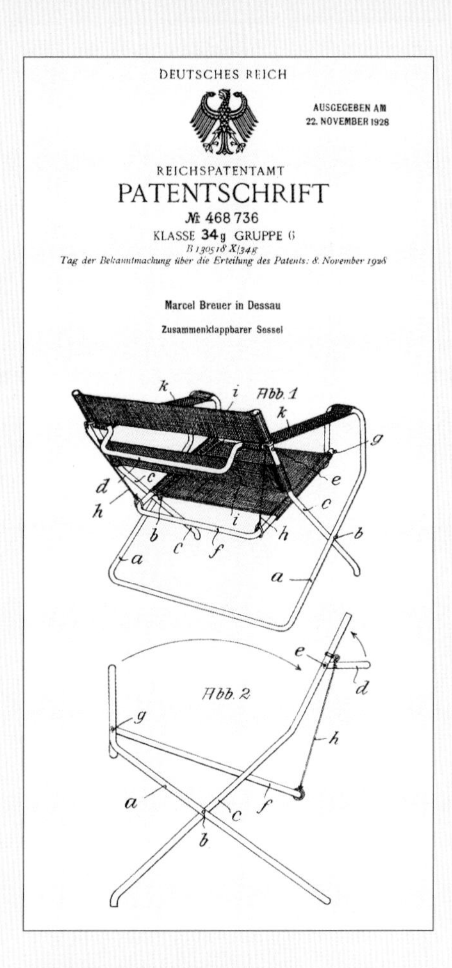
DEUTSCHES REICH

AUSGEGEBEN AM
22. NOVEMBER 1928

REICHSPATENTAMT

PATENTSCHRIFT

№ 468 736

KLASSE 34g GRUPPE 6

B 130518 X/34g

Tag der Bekanntmachung über die Erteilung des Patents: 8. November 1928

Marcel Breuer in Dessau

Zusammenklappbarer Sessel

Marcel Breuer received a *Deutsches Reich Patent* in 1928 for his *Zuzammenklappbarer Sessel*, a collapsible easy chair with steel frame and seat, back and armrests of *Eisengarn*, literally iron yarn. In appearance it is a development of Breuer's *Wassily* chair of 1926. Manufactured by Tecta, Germany.

How many bends in a metal tube does it take to make a good looking and reasonably comfortable chair? Erik Magnussen's steel and canvas designs of 1965 offer two answers. Manufactured by KEVI, Denmark.

The butterfly chair was reportedly invented around 1870 and was apparently used by Italian officers in the 1930s in Ethiopia. It is also called the *Tripoli* chair.

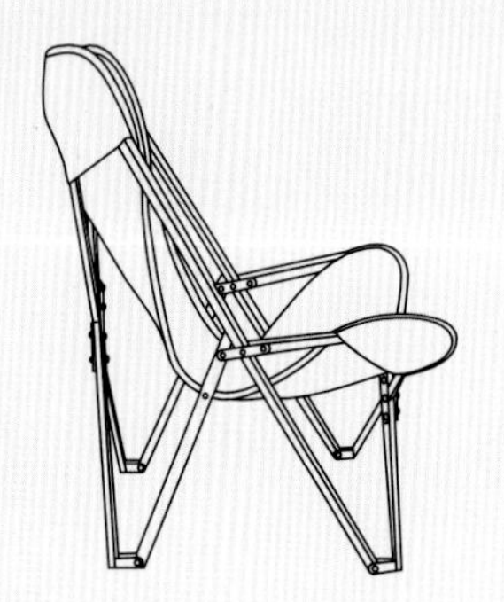

One contemporary version of the Tripoli chair is marketed as the *Bush* chair by the Conran Shop.

Designers Jorge Ferrari-Hardoy, Antonio Bonet and Juan Kurchan redesigned the butterfly chair in 1940. After World War II Knoll International took up production, calling it the *Hardoy* chair. Knoll later lost a law suit attempting to protect their copyright. The chair was already familiar in a number of versions. Manufactured by Knoll International, USA.

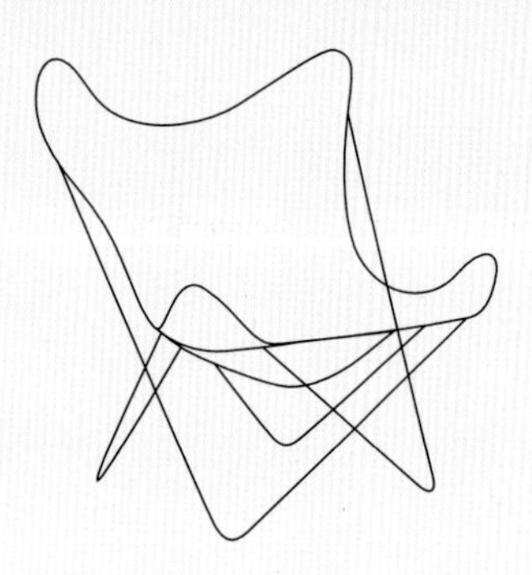

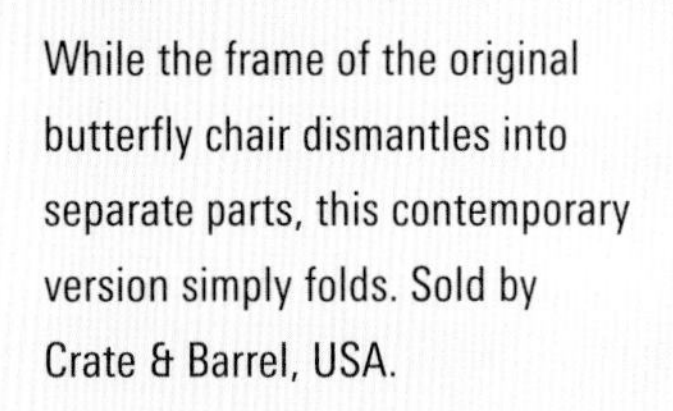

While the frame of the original butterfly chair dismantles into separate parts, this contemporary version simply folds. Sold by Crate & Barrel, USA.

Stacking chairs are cooperative collapsibles. Individual chairs may not fold, but two chairs stacked together occupy less than double the practical space of one. As a group, therefore, they can do what the individual member cannot; multiplicity creates advantages of scale.

Horizontal stackability facilitates transport by bicycle of a hundred Chinese restaurant chairs.

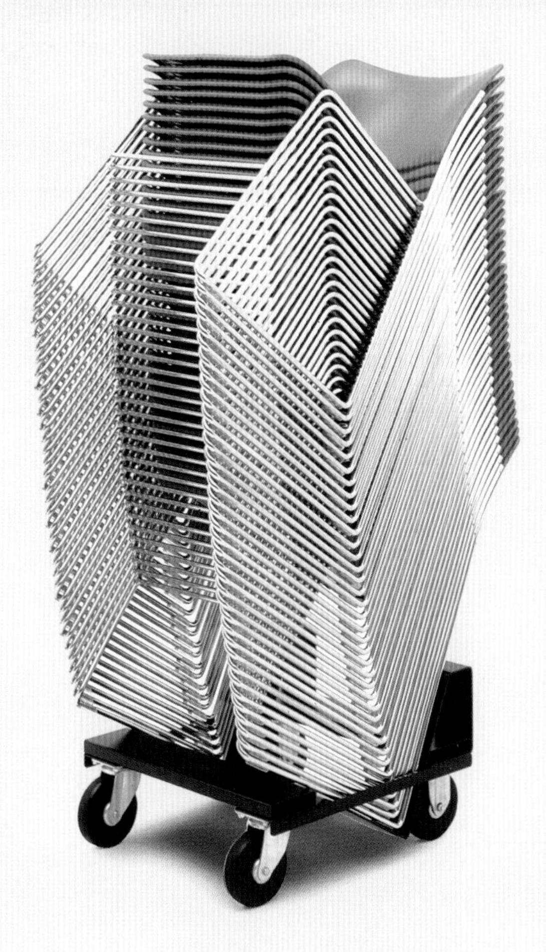

The *GF* chair designed by David Rowland in 1964 is the world champion of stackability. Forty chairs form a tower of just 120 cm (3 ft 11 in). The seat and back of vinyl-covered steel offer a reasonable measure of comfort. Manufactured by GF Office Furniture, USA.

◄
To stack and unstack chairs in a jiffy is one of the first disciplines entrusted to a restaurant waiter.

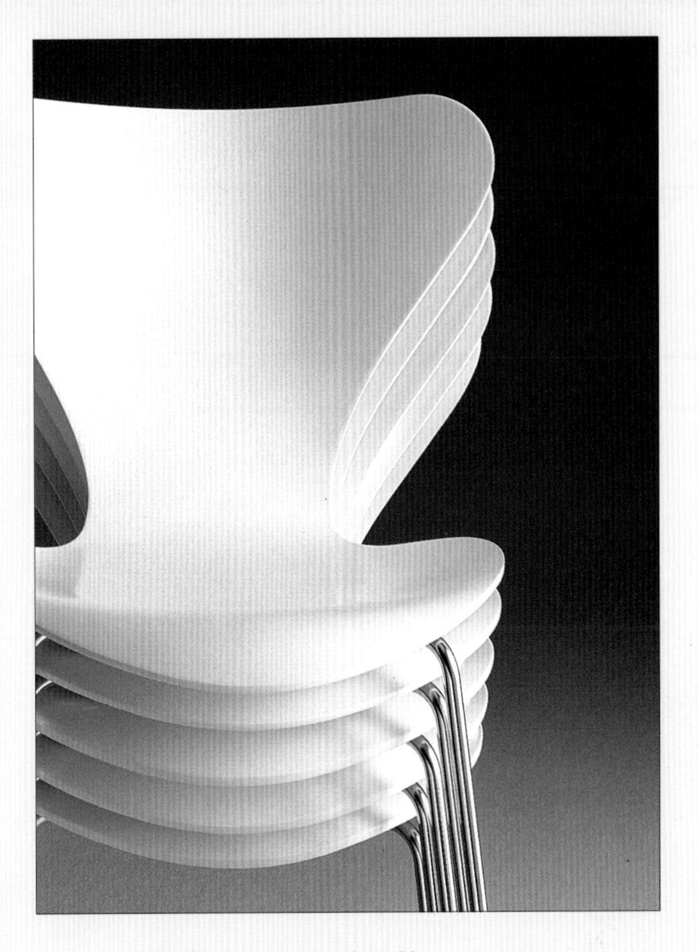

The GF chair may be unbeatable in stackability, but other chairs win out on aesthetics. Arne Jacobsen's *3107* chair from 1954 proves the point. Manufactured by Fritz Hansen, Denmark.

Erik Magnussen's *Chairik* from 1999 is a recent entry in the never-ending contest to combine the greatest comfort, aesthetics and stackability in one affordable chair. Manufactured by Engelbrecht, Denmark.

Louis XX designed by Philippe Starck in 1991 is one chair and two sculptures: one sculpture when the chair stands alone; another when twelve are stacked together. Each chair is made of one metal part and one plastic part. Manufactured by Vitra, Germany.

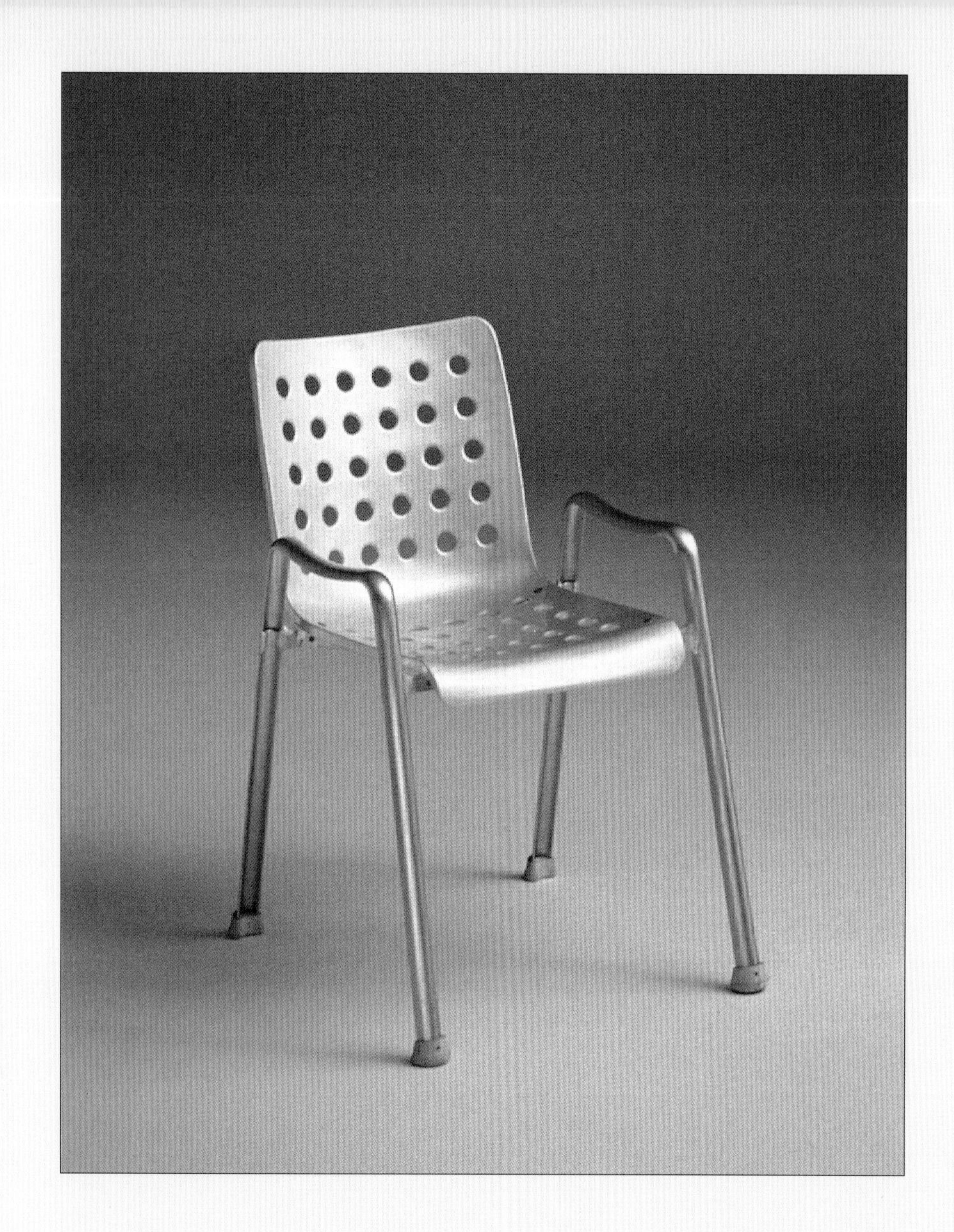

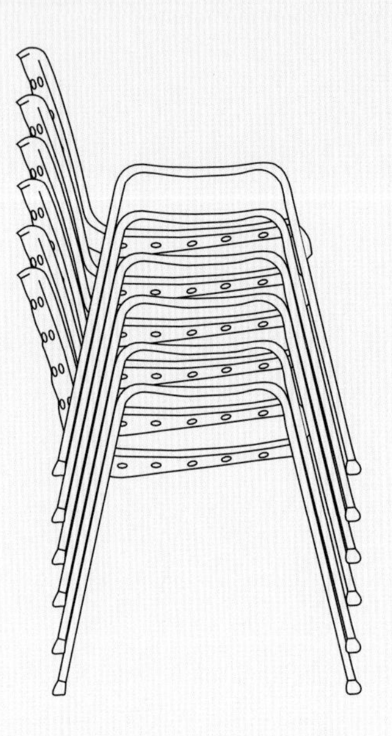

The *Landi* chair was one of the first shell chairs, with seat and back in one piece. It was designed by the painter Hans Coray for the Swiss National Exhibition in 1938. The Landi has been in continuous manufacture for more than sixty years. The holes in the pressed aluminium seat/back are added for the sake of appearence, weight reduction and outdoor utility. The current manufacturer, Zanotta in Italy, calls Landi *Spartana*.

Stacking chairs generally share a common problem: when stacked, they waste a lot of practical space between the legs of the lowest chair. *Santachair* designed by Denis Santachiara avoids that flaw. Stacks of Santachairs, especially small stacks, occupy less practical space than equivalent stacks of other chairs. Santachair has legs and armrests of aluminium, while seat and back are made of plastic. Manufactured by Vitra, Switzerland.

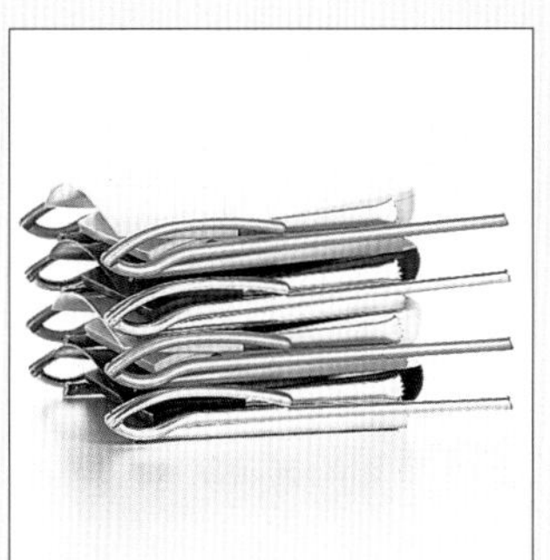

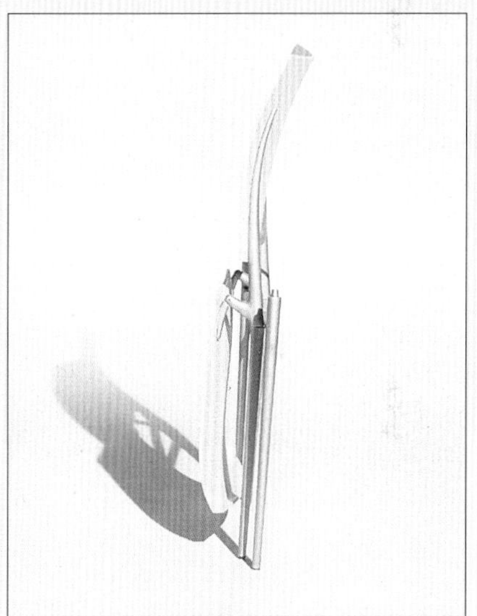

Miss C.O.C.O. designed by Philippe Starck in 1999 is another unexpected candidate for space-saving collapsibility. Manufactured by Cassina, Italy.

Theatres usually have collapsible seats to facilitate fast seating, fast departing and fast cleaning. The result is not always a pleasing view: too much of a good thing is not necessarily beautiful. Sometimes, however, the rows of seats take on sculptural form. In the Bellevue Theatre north of Copenhagen, designed by Arne Jacobsen in 1934–35, more really is more.

Concerts are occasionally played to a less than full house. If the chairs do not have the same acoustic properties when empty as when occupied, then the sound will change from one performance to another. Poul Kjærholm addressed this problem with the chairs he designed for the concert room at the Louisiana art museum, north of Copenhagen, in 1976. Manufactured by PP Møbler, Denmark.

Poul Kjærholm also designed chairs with 'human' acoustic properties in 1978 for Østre Gasværk, an old gasometer converted into a theatre. With these chairs the auditorium would have had the same acoustic properties with or without audience; unfortunately they were never installed. Manufactured by PP Møbler, Denmark.

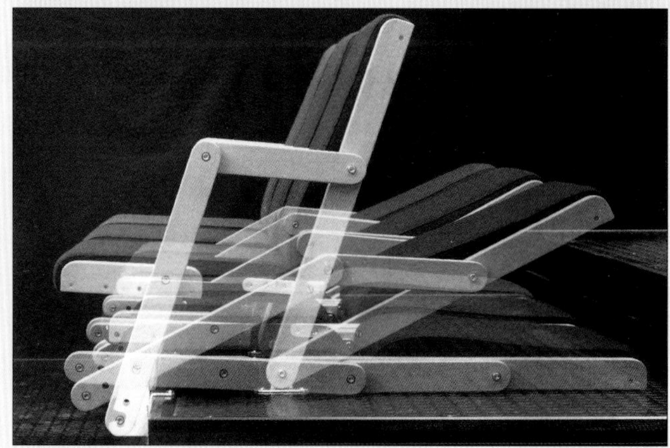

With Bernt's system of theatre seating designed in 1985 for the Ishøj Theater, south of Copenhagen, the vanishing tricks venture off the stage and into the auditorium. After curtain fall, Bernt's chairs collapse and slide under the floor, which is comprised of a number of boxes that finally slide under each other. Manufactured by BCI, Denmark.

Beds, sofabeds and sofas

Stools and chairs are the biggest category of collapsible furniture, maybe because of need, maybe because their form encourages it. But other types of collapsible furniture do exist. Guest beds are a special category —a staple of slapstick comedy, and source of many funny and not-so-funny memories. Today, blow-up collapsibles compete with older designs such as the concertina bed, an alternative to Procrustes' grim strategy of collapsing the person to fit the furniture. A special variety of hidden-by-day sleeping furniture is the sofabed, which folds up as a sofa and folds out to double up as a bed. With a few notable exceptions—such as Osvaldo Borsini's famous design—sofabeds are generally ugly to look at and awkward to lie on.

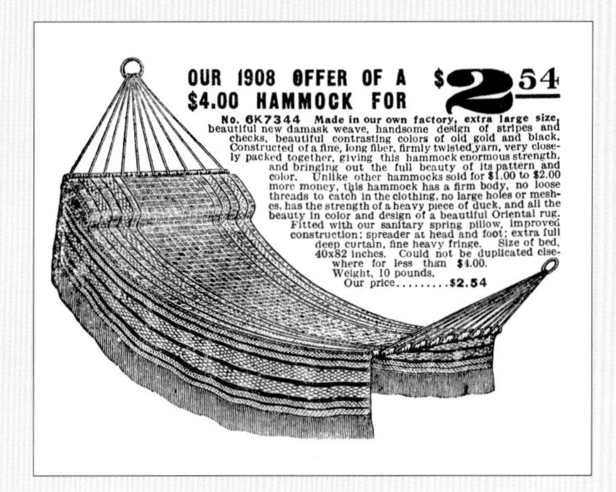

Hammock from the 1908 Sears, Robuck & Co. catalogue, USA.

When Columbus discovered the Americas, he also discovered the fabric suspended sleeping device known today as a hammock. Hammocks are totally collapsible by folding. When not in use, they stow away and take up very little room.

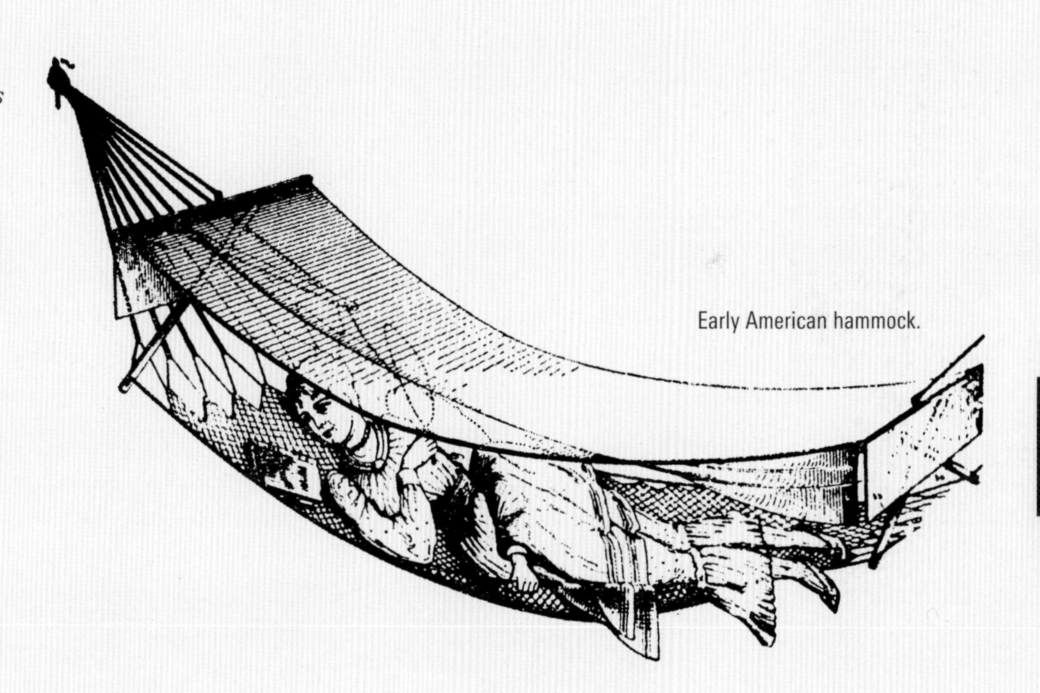

Early American hammock.

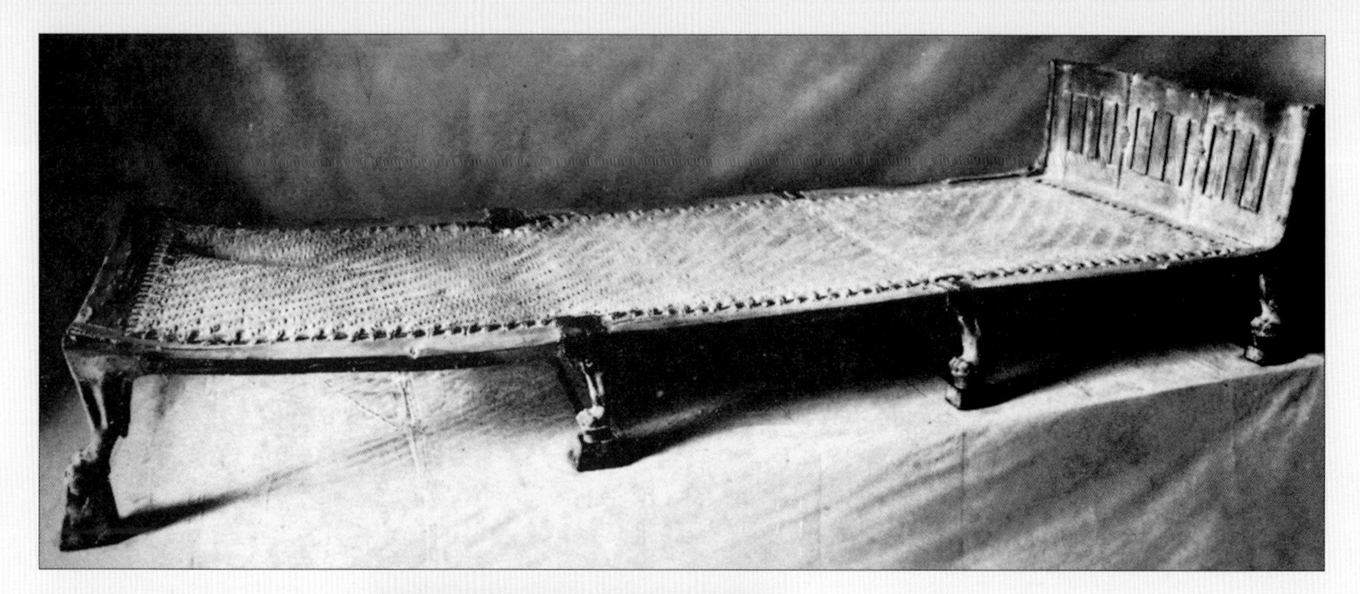

The task of all furniture is to elevate man and his possessions above the ground. The task of the bed is to elevate *resting* man over the ground. The camp bed found in Tutankhamun's tomb suggests that the Egyptian sovereign also demanded comfort and dignity when away from home. The king's bed, made in *c*. 1360 BC, folded on copper hinges.

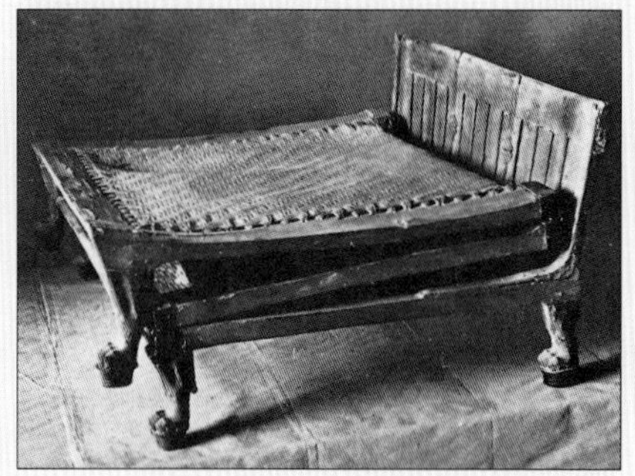

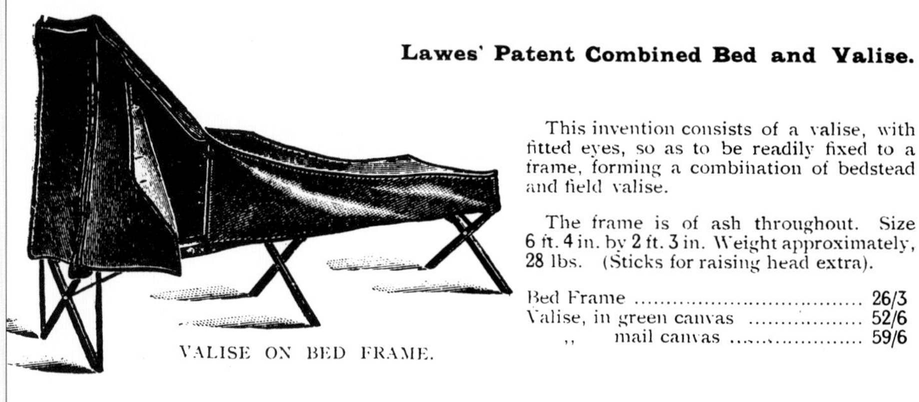

Lawes' Patent Combined Bed and Valise.

This invention consists of a valise, with fitted eyes, so as to be readily fixed to a frame, forming a combination of bedstead and field valise.

The frame is of ash throughout. Size 6 ft. 4 in. by 2 ft. 3 in. Weight approximately, 28 lbs. (Sticks for raising head extra).

Bed Frame 26/3
Valise, in green canvas 52/6
,, mail canvas 59/6

VALISE ON BED FRAME.

Lawes' Patent Combined Bed and Valise. Army & Navy Stores catalogue, 1907, UK.

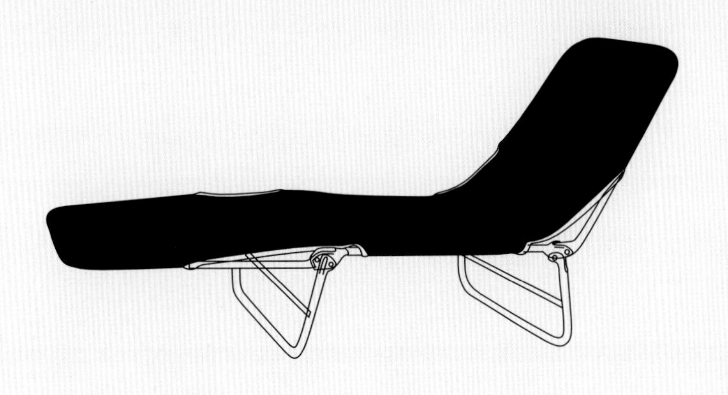

When there are no ship's masts or trees around, other instant resting devices take over from the hammock. The so-called *Dream Bed* from the 1950s has for sure induced many bad dreams.

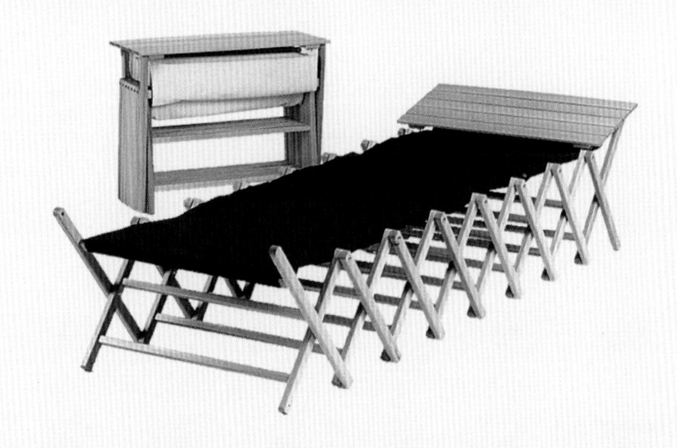

The *Concertina* bed is yet another anonymous design that has recently been taken up for redesign – and repricing. By day it's a small table, by night a bed for one not-too-grown-up person. Manufactured by Hyllinge Træindustri, Denmark.

Ole Gjerløv Knudsen's brilliant camp bed from 1964 is held together by a bucksaw principle, like his bucksaw chair (p. 185). When folded it is wrapped in its own cloth. Manufactured by Trip-Trap, Denmark.

The bed-disguised-as-wardrobe theme was explored by the Shakers and after them by Charles Eames. Børge Mogensen and Grethe Meyer took it up once again in 1957. Their wardrobe front is of Oregon pine. Manufactured by Boligens Byggeskabe, Denmark.

Osvaldo Borsini's design from 1954 is an all-time winner among sofabeds for both aesthetics and fast functionality. Manufactured by Tecno, Italy.

Camilla by Achille Castiglioni and Giancarlo Pozzi, an anonymous folding seat reinvented for the design market in 1984. Manufactured by Zanotta, Italy.

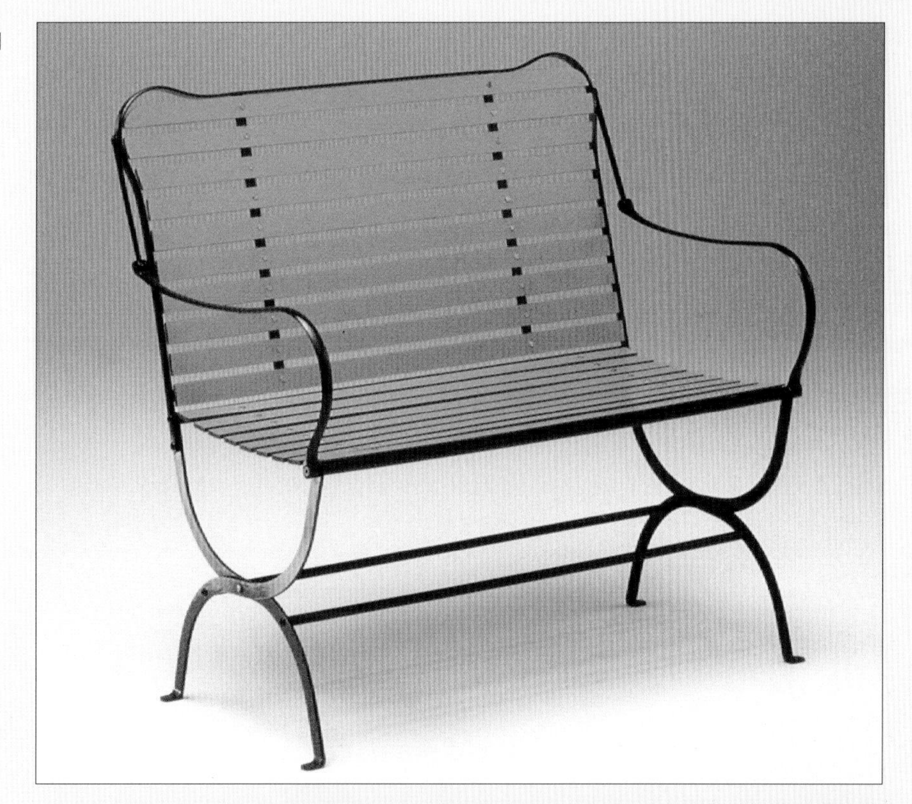

Stefan Heiliger designed *Bat,* a truly collapsible sofa, in 1986. It comes in many colours with covers of leather, canvas or nylon. Manufactured by Strässle, Switzerland.

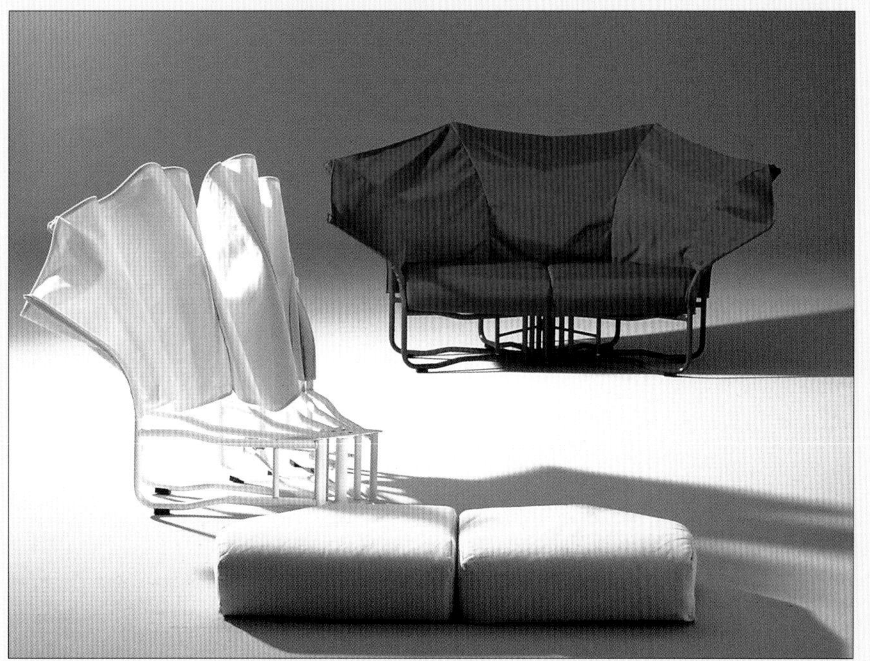

Tables

Two conditions must be met before collapsible objects are designed and manufactured: there must be a demand and there must be a technical possibility. Tables meet both requirements: many of us have an occasional need for the elevated surface space of a table; and collapsibility is also technically possible. There are two degrees of collapsibility of tables. Some dismantle and fold away completely; others expand and reduce only partially by means of extension plates. There are also tables that collapse collectively. Some – usually of uniform size – stack; others – in sets of decreasing size – nest.

This drop-leaf Shaker table of cherry wood, made in *c.* 1825–50 in New Lebanon, New York, represents thousands of anonymous tables of the same type, with or without a drawer.

Charles Rennie Mackintosh designed his black drop-leaf table as part of a dining set for B. Lowke in 1918. Manufactured by Cassina, Italy.

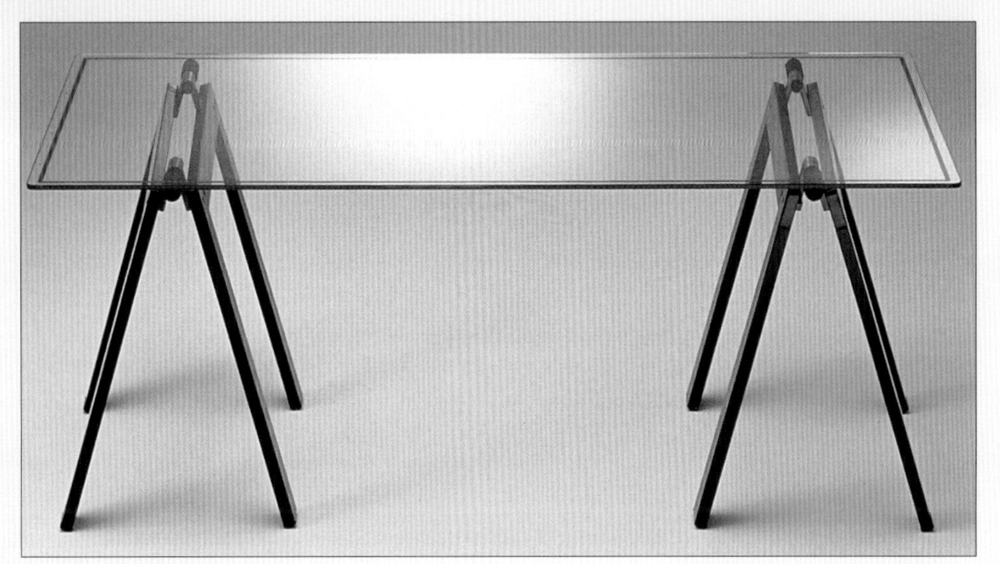

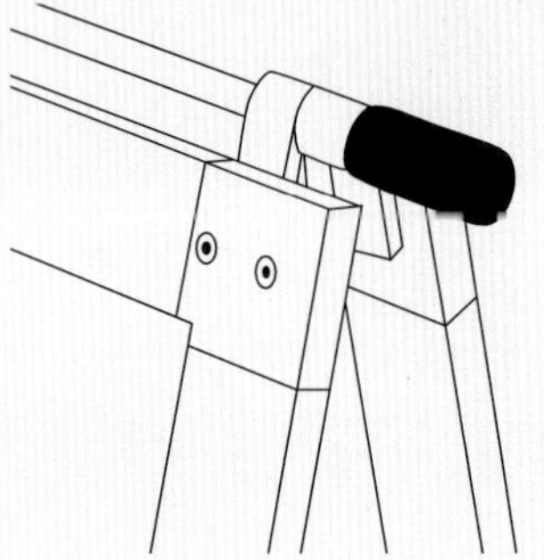

Gaetano table on collapsible trestles designed by Gae Aullenti in 1974. Manufactured by Zanotta, Italy.

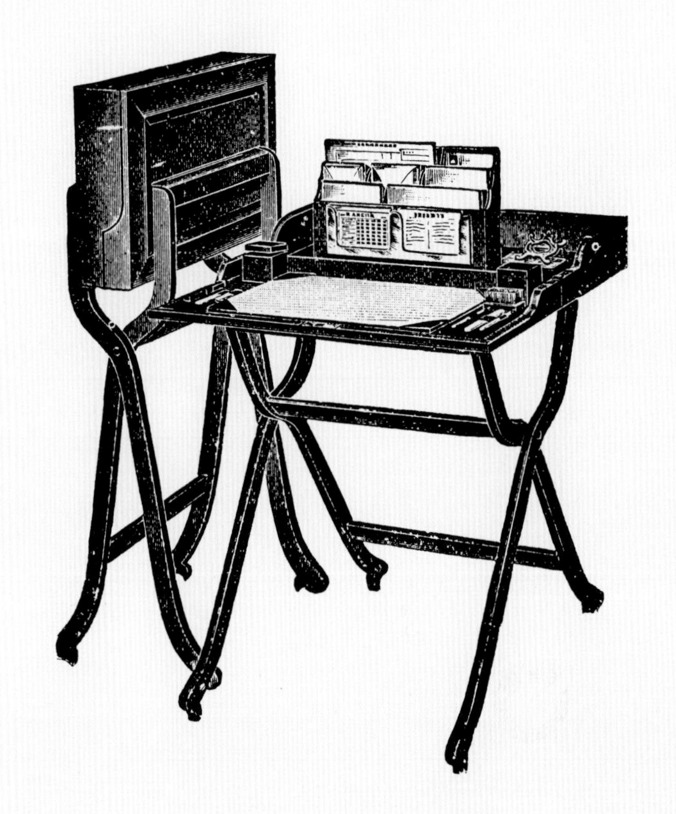

Folding garden table, and *Princess* folding writing table of oak or mahogany. Army & Navy Stores catalogue, 1907, UK.

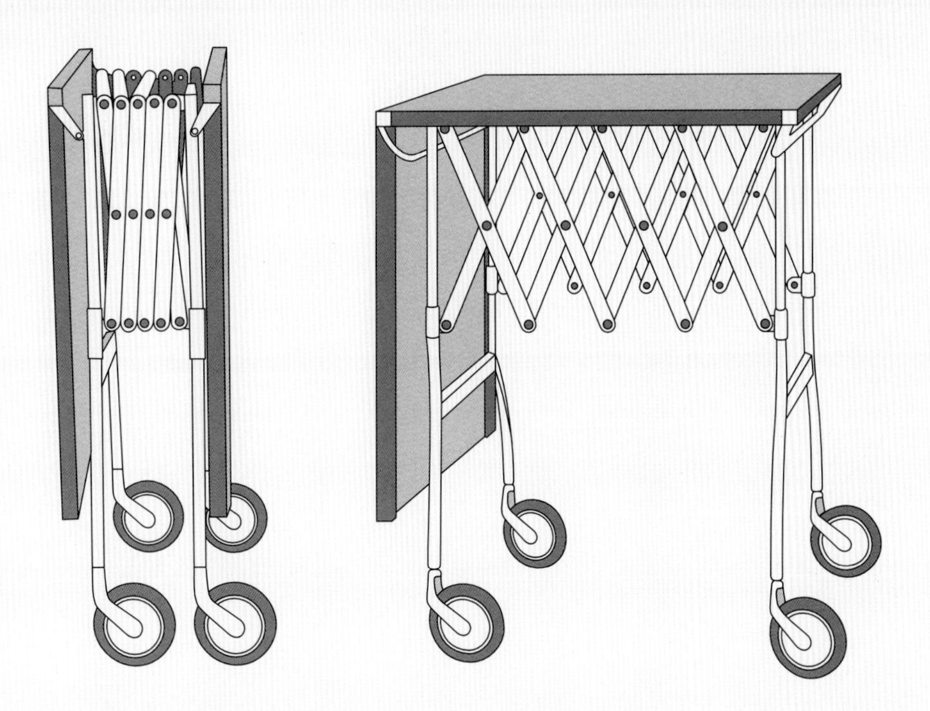

Battista, a length-adjustable, concertina-collapsible table designed by Antonio Citterio and Oliver Löw. Manufactured by Kartell, Italy.

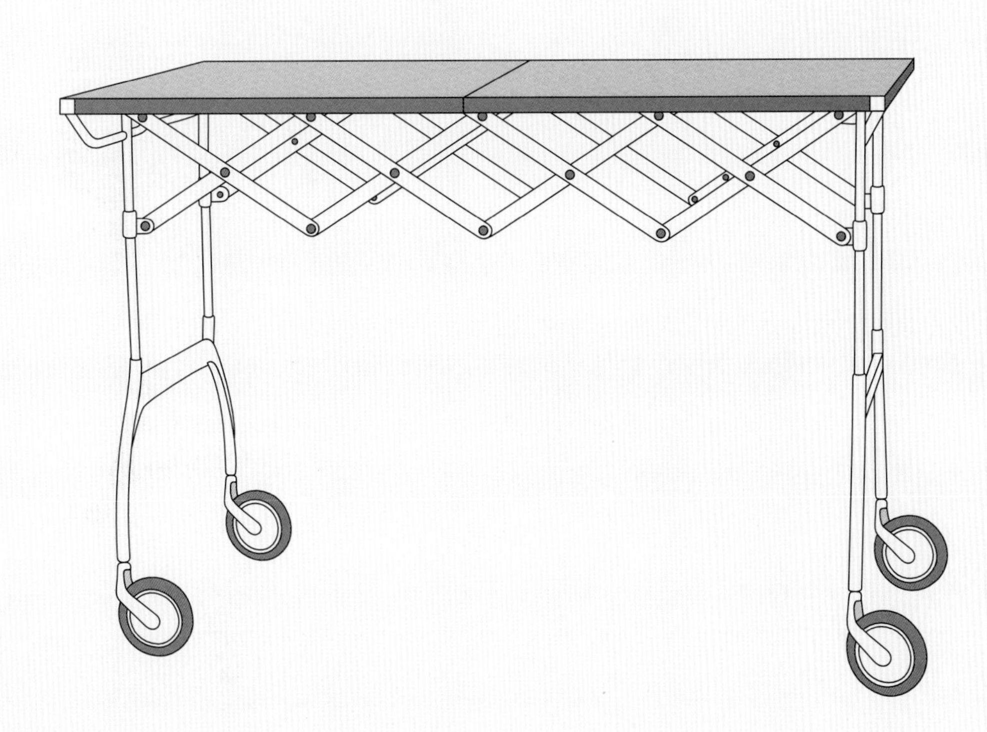

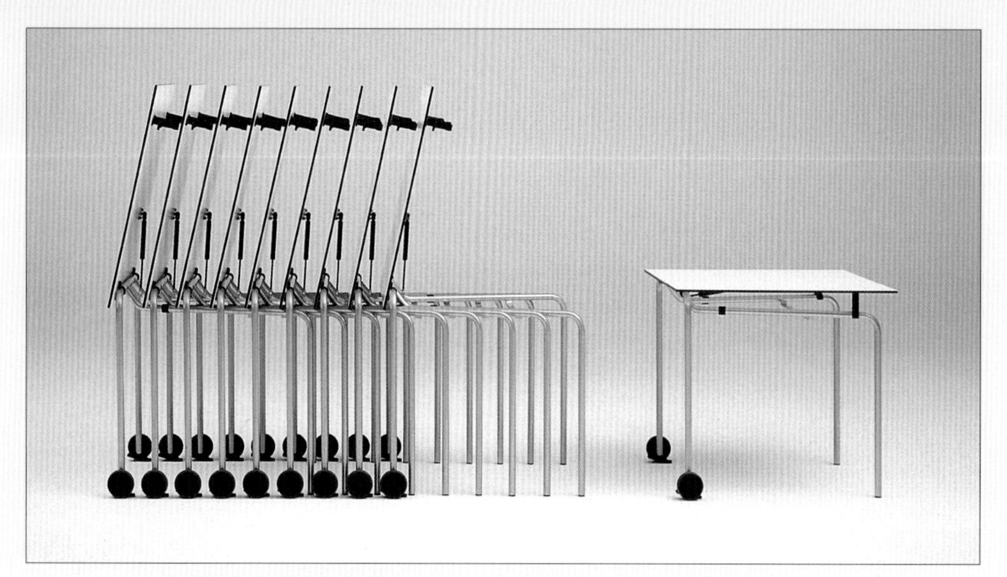

Move It – horizontally stacking tables designed by Alfredo Walter Häberli and Christophe Marchand. Manufactured by Thonet, Germany.

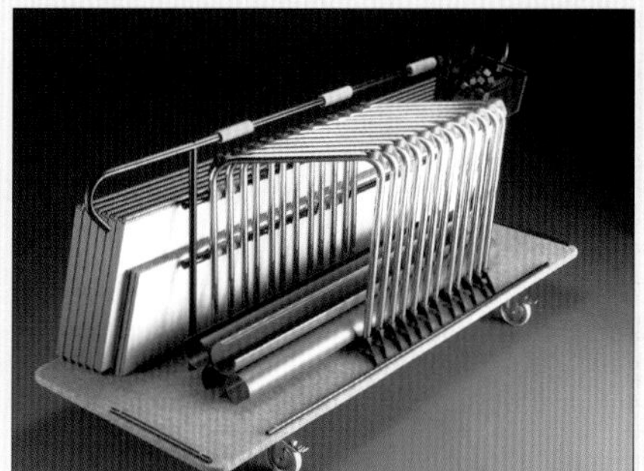

Click – easy-to-assemble tables that dismantle and stow away on a purpose-built trolley. Designed by Erik Magnussen in 1995. Manufactured by Fritz Hansen, Denmark.

Rolled up and folded away in its sack, this hardwood table measures 110 x 20 cm (3 ft 7 in x 8 in). Rolled out for a picnic or any other temporary purpose, it presents a top of 108 x 76 cm (3 ft 6 in x 2 ft 6 in). Company design, *c*. 1996. Manufactured by Byers of Maine, USA.

Mogens Koch designed this table for the Danish embassy in Washington DC. The frame is beech and the top teak. It goes with his folding chair (p. 180). Manufactured by Rud. Rasmussens Snedkerier, Denmark.

The folding secret of *La Loggia* by Mario Bellini is hidden under the table top. Manufactured by Cassina, Italy.

La Barca designed by Piero de Martini is a half-way solution. Both top and base are hinged, allowing the table to halve in size. Manufactured by Cassina, Italy.

Luncheon Table and Seats. Army & Navy Stores catalogue, 1907, UK.

Simsalabim (the Swedish equivalent of 'abracadabra') was designed by Börge Lindau. One table folds to half the size; two halved tables occupy the same space as one. Magic! Manufactured by Blå Station, Sweden.

No flaps, one flap, two flaps. Designed by Ludwig Roner and manufactured by Wogg, Switzerland.

No flaps, one flap, two flaps. Designed by Niels Jørgen Haugesen, 1986. Manufactured by Fritz Hansen, Denmark.

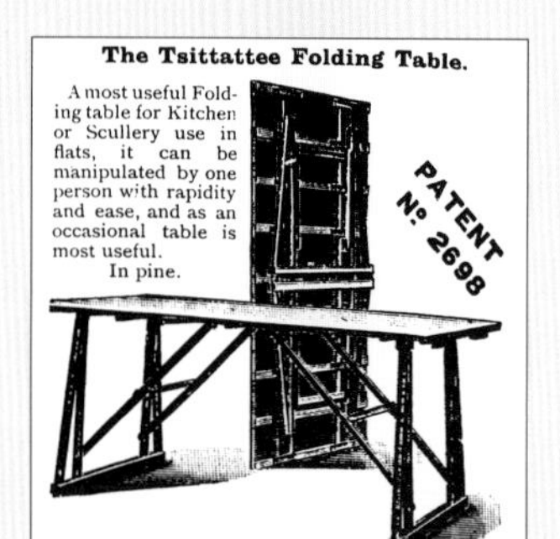

The *Tsittattee Folding Table*. Army & Navy Stores catalogue, 1907, UK.

Broomstick table by Vico Magistretti, 1978–79. Manufactured by Alias, Italy.

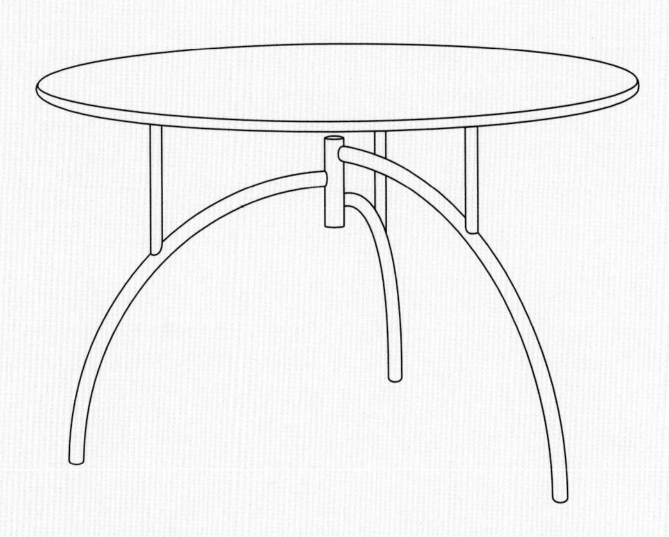

Guess how it folds: *Tippy Jackson* by Philippe Starck, 1982. Manufactured by Driade, Italy.

Kaare Klint and Joe Colombo both had a lucky hand when designing their card tables. Klint focused on reduction (left); Colombo on expansion (below).

Card table by Kaare Klint, 1934. Manufactured by Rud. Rasmussens Snedkerier, Denmark.

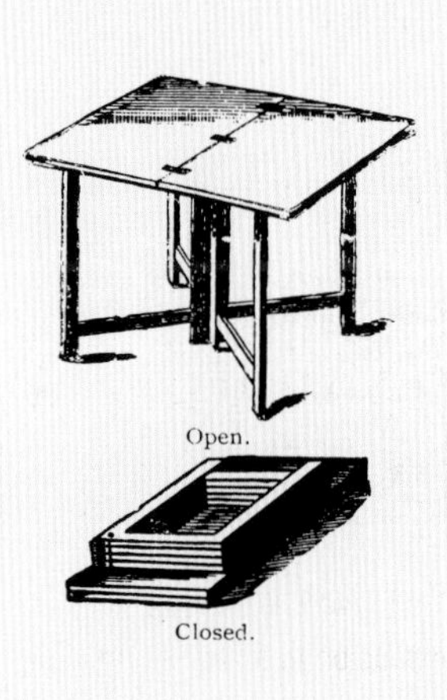

Poker Table by Joe Colombo, 1968. Manufactured by Zanotta, Italy.

The *Lennox* portable table, patented by Lieutenant-General Sir Wilbraham Lennox. Army & Navy Stores catalogue, 1907, UK.

Poul Kjærholm did not invent the extension to a circular table top, but his design is very convincing both in terms of storage and use. A ring of solid maple in six pieces does the trick. *PK54*, 1963, manufactured by Fritz Hansen, Denmark.

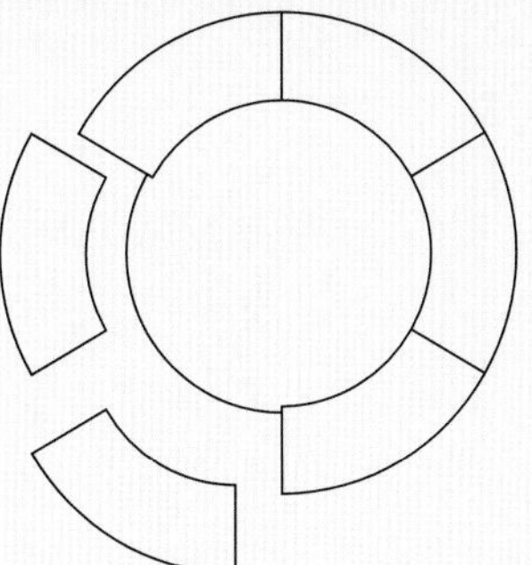

Bernt squared the circle in 1985. Manufactured by Carl Hansen, Denmark.

Real designers are full-time designers. They never stop thinking of ways to improve their surroundings. Given the number of coffee tables they have come up with, you would be forgiven for concluding that they spend most of their time taking breaks.

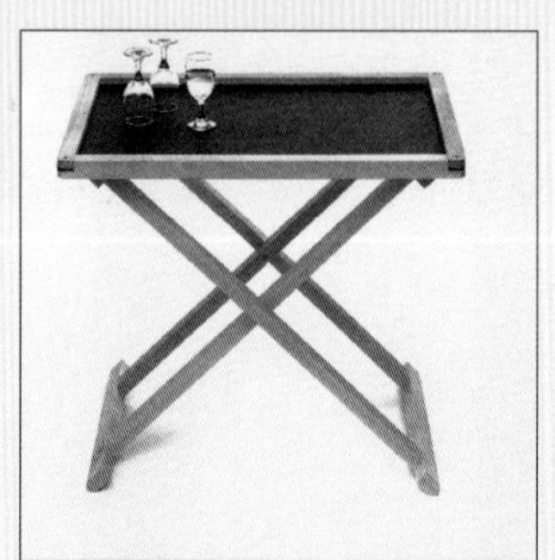

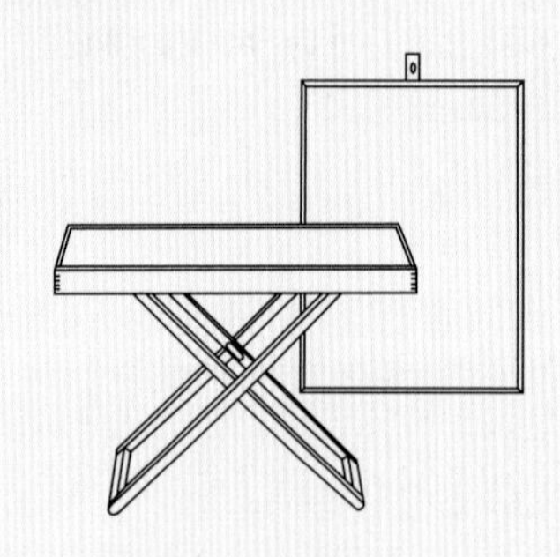

Tray table by Hans Bølling. Manufactured by Torben Ørskov, Denmark.

Collapsible table by Torben Lind, 1962. Manufactured by Torben Ørskov, Denmark.

Collapsible table by Jørgen Møller. Manufactured by Torben Ørskov, Denmark.

Cumano occasional table by Achille Castiglioni, 1979. Manufactured by Zanotta, Italy.

Serafino folding table by Emaf Progretti, 1986. Manufactured by Zanotta, Italy.

Cipango folding table. Manufactured by Zanotta, Italy.

Tray table by Hans J. Wegner. Manufactured by PP Møbler, Denmark.

Ship table by Kaare Klint, 1933. Manufactured by Rud. Rasmussen, Denmark.

Tray table by Mogens Koch, 1960. Manufactured by Rud. Rasmussen, Denmark.

Nests of tables are clearly an irresistible challenge to designers.

▶

Marcel Breuer, 1926.

▶▶

Alvar Aalto, 1933.
Manufactured by Artek, Finland.

Dieter Rams. Manufactured by Vitsoe, Germany.

Poul Kjærholm, 1957.
Manufactured by Fritz Hansen, Denmark.

Afra & Tobia Scarpa.
Manufactured by Cassina, Italy.

Bernt, 1985.
Manufactured by Carl Hansen & Søn, Denmark.

Storage furniture

Collapsible storage furniture is a sparsely populated category. This is perfectly logical, for the function of storage furniture is to hold objects – unlike stools, chairs, beds and sofas, which hold bodies. Man's need for containers for his possessions is not as variable as his need for body furniture. We simply do not want to fold a cupboard. Where would we keep its contents if we did? Professional travellers tend to travel light. When they can't, they will stow their belongings in large suitcases which function as furniture.

Vico Magistretti made two highly original collapsible bookshelves. In 1978–79, he designed the *Broomstick* range comprising tables, hat stand and bookshelf, everything absolutely collapsible. Manufactured by Alias, Italy.

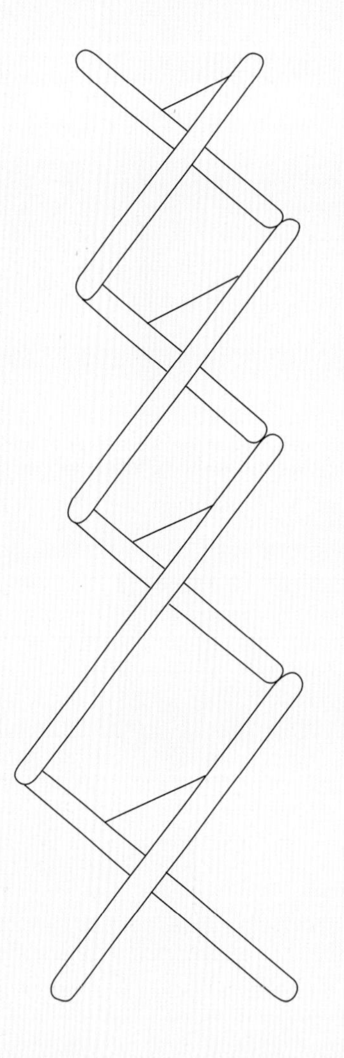

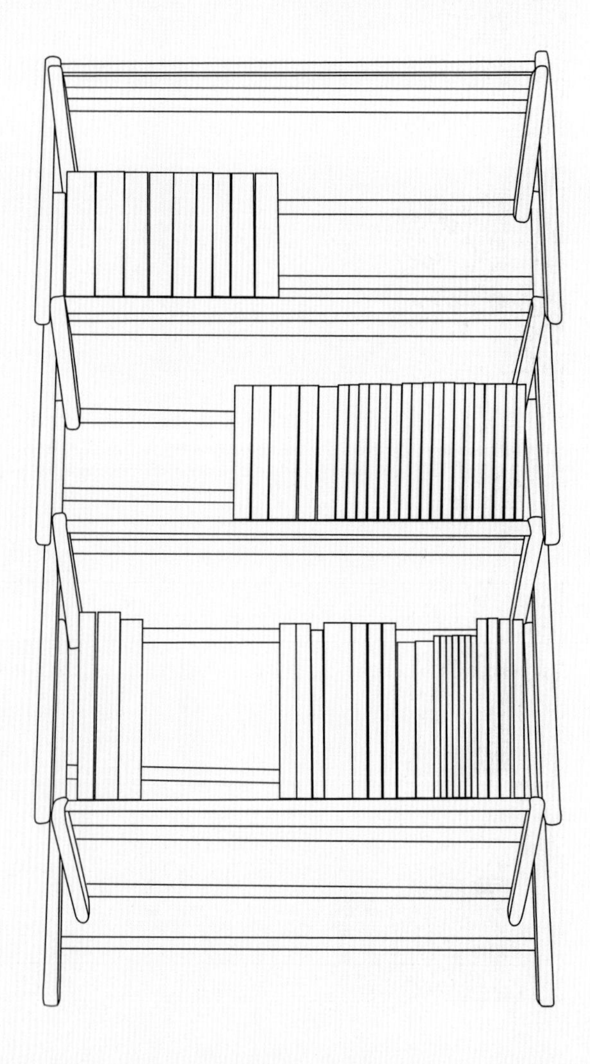

In 1977 Magistretti designed the *Nuvola Rossa* ('red cloud') shelving system. These are definitely not bookshelves for a public library, but their nomadic quality is without parallel. A few turns with an allen key and your red cloud is ready to drift. *Nuvola Rossa* comes – despite the name – in natural beech and black. Manufactured by Cassina, Italy.

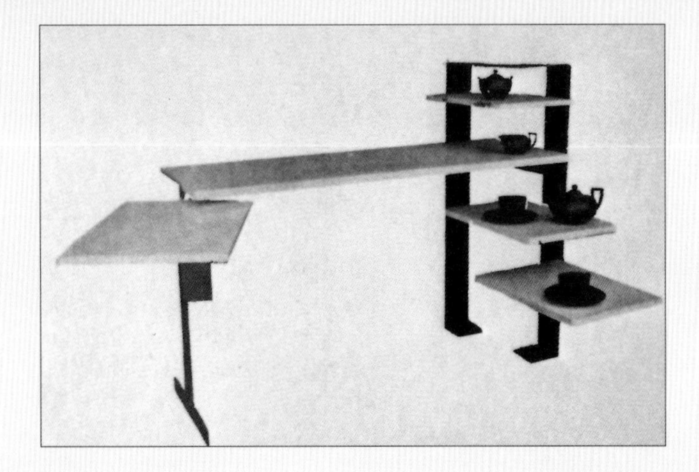

▲

A now-forgotten designer conceived this excellent system in 1929.

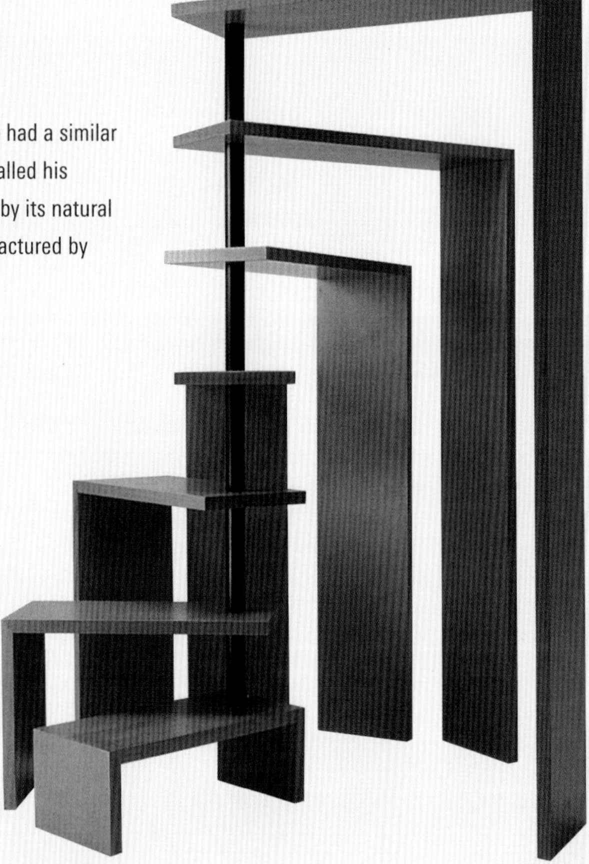

►

Achille Castiglione had a similar idea in 1989. He called his rotating shelf unit by its natural name: *Joy*. Manufactured by Zanotta, Italy.

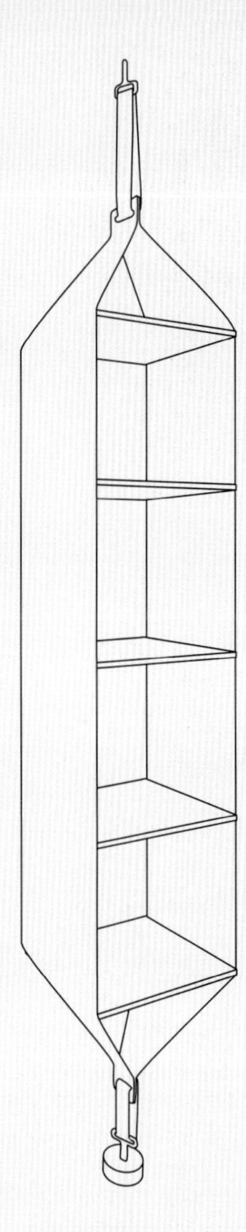

Suspended canvas bookshelf with shelves of aluminium plates, designed by Jørgen Høj.

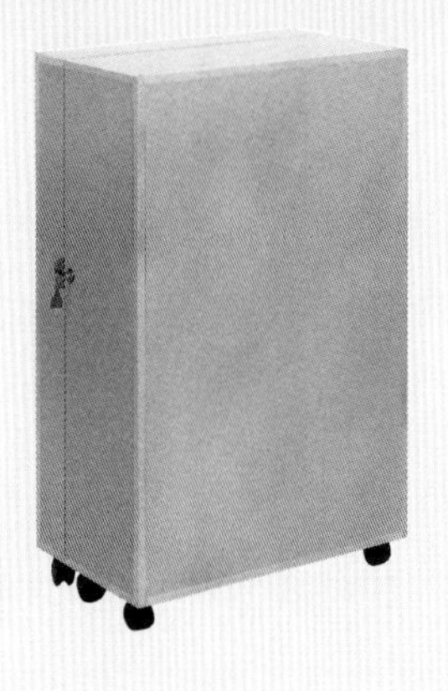

As more and more people stay at home to do more and more work on computer, their furniture requirements change. *Post-Box* is a serious contribution to the home office. It spreads out to provide a work surface and folds away again for space and privacy. Manufactured by Åhmans i Åhus, Sweden.

The trunk cupboard by Hans Eichenberger, 1983, combines mobility and fold-out quality. Manufactured by Wogg, Switzerland.

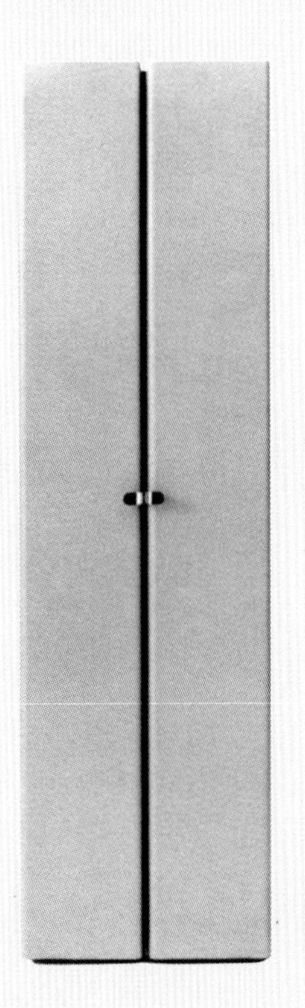

Sail away!

The world is in flux. Life is a voyage. The circumstances under which man lives change, and so do his needs and wants. Collapsibles are man's adaptation to that changing world. Their elegant efficiency enhances his life – your life – on the move. Bon voyage!

Alison, Filippo
Charles Rennie Mackintosh as a Designer of Chairs, 1973
Documenti di Casabella I / Warehouse Publications, UK

Andersen, Rigmor
Kaare Klint Møbler, 1979
Kunstakademiet, Denmark

Aubury, Edward
The Bailey Bridge and Uniflote Handbook, 1962
Acrow Press, UK

Bénaïm, Laurence
Issey Miyake, 1997
Universe Publishing, US

Bill, Max
Form, 1952
Verlag Karl Werner, Switzerland

Bottomley, I and A.P. Hopson
Arms and Armor of the Samurai, 1988
Crescent Books, US

Buchet, Martine
Panama: A Legendary Hat, n.d.
Editions Assouline, France

Casciani, Stefano
Furniture as Architecture, 1988
Arcadia Edizioni, Italy

Chapuis, Bernard & Ermine Herscher
Qualités: Objets d'en France, 1997
Éditions Du May, France

Conran, Terence
Design, 1996
Conran Octopus, UK

Cornfeld, Betty & Owen Edwards
Quintessence: The Quality of Having It, 1983
Crown Publishers, US

Design Mirroir du Siècle, 1993
Flammarion, France

Design Dasein
Museum für Kunst und Gewerbe Hamburg 1987, Germany

Doblin, Jay
One Hundred Great Product Designs, 1970
Van Nostrand Reinhold, US

Eden, Mary & Richard Carrington
Sengen, 1961
Samlerens Forlag, Denmark

Faegre, Torvald
Tents: Architecture of the Nomads, 1979
Anchor Books, US

Forty, Adrian
Objects of Desire, 1986
Thames & Hudson, UK

Frey, Gilbert
The Modern Chair: 1850 to Today, 1970
Arthur Niggli

Gandy, Charles D. & Susan Zimmermann-Stidham
Contemporary Classics: Furniture of the Masters, 1981
McGraw-Hill, US

Geest, Jan van & Otakar Macel
Stühle aus Stahl, 1980
Walther König, Germany

Harlang, Christoffer, Keld Helmer-Petersen & Krestine Kjærholm (eds.)
Poul Kjærholm, 1999
Arkitektens Forlag, Denmark

Harpur, Brian
A Bridge to Victory, 1991
Ministry of Defence/HMSO, UK

Hatton, E.M.
The Tent Book, 1979
Houghton Mifflin, US

Heal's Catalogue 1853–1934
Middle Class Furnishing, 1972
David & Charles, UK

Hennessey, James & Victor Papanek
Nomadic Furniture, 1973
Pantheon Books, US

Herløw, Erik
Gode ting til hverdagsbrug, 1949
Det Schønbergske Forlag, Denmark

Hillman, David & David Gibbs
Century Makers, 1998
Weidenfeld & Nicolson, UK

Jepsen, Anton
Danske snedkermøbler gennem 125 år, 1994
Rud. Rasmussens Snedkerier, Denmark

Johnson, Martin
Safari, 1928
Gyldendal, Denmark

Karlsen, Arne
Dansk Møbelkunst i det 20. århundrede, 1990
Christian Ejlers, Denmark

Karlsen, Arne
Møbler tegnet af Børge Mogensen, 1968
Arkitektens Forlag, Denmark

Katz, Sylvia
Classic Plastic, 1984
Thames and Hudson, UK

King, Geoffrey
Miniature Antique Maps, 1997
Map Collector Publications, UK

Langkilde, Hans Erling
Pæne ting ombord, 1944
Emil Wienes Forlag, Denmark

Levine, Bernard
Pocket Knives, 1998
Chartwell Books, US

Logan, William Bryant
The Tool Book for the Well-Tended Garden, 1997
Workman, US

Manasek, F.J.
Collecting Old Maps, 1998
Terra Nova Press, US

Mang, Karl
History of Modern Furniture, 1979
Academy Editions, UK

Manzini, Ezio
The Material of Invention, 1986
Arcadia Edizioni, Italy

McDermott, Catherine
Book of 20th Century Design, 1998
The Overlook Press, US

Mollerup, Per
Design att leva med, 1997
Carlsson, Sweden /
Design er ikke noe i sig selv, 1998
Norsk Form, Norway /
Design er ikke noget i sig selv, 1998
Gyldendal, Denmark

Mollerup, Per
Jørgen Gammelgaard, exhib. cat., 1995
The Danish Museum of Decorative Art, Denmark

Mollerup, Per
Poul Kjærholm's Furniture, 1981
Mobilia Press, Denmark

Mouret, Jean-Noël
Knives of the World, 1995
Bramley Books, UK

Myrdal, Jan
En Meccano pojke berättar, 1989
Wiken, Sweden

Møbeltegninger, 1984
Erhvervsskolernes Forlag, Denmark

Møller, Svend Erik (ed.)
På Wegners tid, 1989
Poul Kristensens Forlag, Denmark

Møller, Henrik Sten
Dansk Design / Danish Design, 1975
Rhodos, Denmark

Møller, Henrik Sten
Tema med variationer, 1979
Sønderjyllands Kunstmuseum, Denmark

Møller, Henrik Sten
Erik Magnussen, 1990
Rhodos, Denmark

Nelson, George
How to See, 1977
Little, Brown, US

Nielsen, Johan Møller
Wegner, en dansk møbelkunstner, 1965
Gyldendal, Denmark

Oda, Noritsugo
Danish Chairs, 1996
Korinsha Press (JP)

Olmert, Michael
Milton's Teeth & Ovid's Umbrella, 1996
Simon & Schuster, US

Packaging bags to trunks, 1994
Thames & Hudson, UK

Pallingston, Jessica
Lipstick, 1999
St. Martin's Press, US

Payne, Christopher (ed.)
Sotheby's Concise Encyclopedia of Furniture, 1989
Conran Octopus, UK

Pearce, Chris,
Twentieth Century Design Classics, 1991
Blossom, UK

Philippi, Simone (ed.)
Starck, 1996
Taschen, Denmark

Poynter, Dan
The Parachute Manual (1991)
Para Publishing, US.

Ragas, Meg Coghen & Karen Kozlowski
Read My Lips: A Cultural History of Lipstick, 1998
Chronicle Books, US

Rasmussen, Steen Eiler
Britisk brugskunst, 1933
The Danish Museum of Decorative Arts, Denmark

Schaefer, Herwin
The Roots of Modern Design, 1970
Studio Vista, UK

Sengen,
The Bed, 1969
Mobilia, Denmark

Skagerfors, Mona
Herrar och den intressanta ytan, 1992
Carlssons, Sweden

Spalt, Johannes (ed.)
Folding Tables, 1987
Birkhäuser, Switzerland

Sprigg, June
Shaker Design, 1986
Whitney Museum of American Art, US

Sprigg, June & David Larkin
Shaker: Life, Work and Art, 1988
Cassell, UK

Sudjic, Deyan
Cult Objects, 1985
Paladin Books, UK

Sugar, Andrew
The Complete Tent Book, 1979
Contemporary, US

Tambini, Michael
The Look of the Century, 1996
Dorling Kindersley, UK

Tenner, Edward
Why Things Bite Back, 1996
Vintage Books, US

Thonet Bentwood & Other Furniture: The 1904 Illustrated Catalogue, 1980
Dover Publications, US

Vries, Leonard de
Victorian Inventions, 1971
John Murray, UK

Wanscher, Ole
Sella Curulis, The Folding Stool: Ancient Symbol of Dignity, 1980
Rosenkilde & Bagger, Denmark

Wanscher, Ole
Møblets Æstetik, 1985
Arkitektens Forlag, Denmark

Wessel, Joan & Nada Westerman
American Design Classics, 1985
Design Publications, US

Wirth, Dick & Jerry Young
Ballooning, 1984
Orbis, UK

Åke Axelson Trettionio Stolar, exhib. cat., 1995
Konstakademien, Sweden

Sources of photos

7, © Disney Enterprises, Inc., California. 10, fire, Polfoto. 13, wickerchair, Woldbye & Klemp. 17, chapeau claque (2), Woldbye & Klemp. 18, elephants, Science Photo Library, London. Photograph by John Reader. 19, venus trap, Biofoto. 23, Dixie Cups, Woldbye & Klemp. 25, deck watch (2), Woldbye & Klemp. 27, drinking glass (2), Woldbye & Klemp. 31, room divider, K. Helmer-Petersen. 32, asparagus, Woldbye & Klemp. 33, reflector, Woldbye & Klemp. 34, bowl, Woldbye & Klemp. 35, Voyager (2), Woldbye & Klemp. 35, wet-weather jacket (2), Woldbye & Klemp. 35, rain hoods, Woldbye & Klemp. 35, plastic sacks, P.A. News, London. Photograph by Adam Butler. 36, napkin, Woldbye & Klemp. 37, Bantry Bay Longboat, Nic Compton, Lewes, East Sussex. 38, moon landing, NASA, Washington. 41, water sack (3), Woldbye & Klemp. 41, watering can (2). Woldbye & Klemp. 42, tote bag, Woldbye & Klemp. 42, string bag, Woldbye & Klemp. 42, plastic bags, Jørgen Schytte. 43, rubber gloves, Woldbye & Klemp. 43, skipping rope, Jeppe Aagaard Andersen. 43, sign, Per Mollerup. 44, black tent, Biofoto. 46, scout camp, Thorbjorn Larsen/KFUM. 50, suspended sail, Jørgen Schytte. 51, circus, Heine Pedersen, Billedhuset. 51, clown, Heine Pedersen, Billedhuset. 52, postcard, Woldbye & Klemp. 53, mug (2), Woldbye & Klemp. 54, paper globe, Woldbye & Klemp. 55, escape maps (2), Woldbye & Klemp. 56–57, survey map (2), Woldbye & Klemp. 58, London Overground/ Underground, Woldbye & Klemp. 58, Flytoget (2), Woldbye & Klemp. 59, Stockholm, Woldbye & Klemp. 59, New York, Woldbye & Klemp. 60, books, Woldbye & Klemp. 61, Baedeker, Woldbye & Klemp. 61, Michelin, Woldbye & Klemp. 62, Bouncing Bugs (2), Woldbye & Klemp. 62, Menneskets Krop, Woldbye & Klemp. 63, newspapers, Jørgen Schytte. 64, paper decorations, Woldbye & Klemp. 65, paper lampshades, Woldbye & Klemp. 65, rugby ball, Woldbye & Klemp. 66, vase, Woldbye & Klemp. 66, Muji storage box, Woldbye & Klemp. 70, airport gate, Per Mollerup. 70, shoe shelf (2), Woldbye & Klemp. 71, bellows lamp (2), Woldbye & Klemp. 72, Imperial camera, Woldbye & Klemp. 74, Markies (4), Eduard Böhtlink. 77, spinosaurus, Woldbye & Klemp. 78, jigsaw puzzle, Woldbye & Klemp. 79, House of Cards, Woldbye & Klemp. 82, shaving gear (2), Woldbye & Klemp. 85, Tuborg bottle, Per Mollerup. 86, crane, Jørgen Schytte. 88, PowerBook, Woldbye & Klemp. 92, umbrella (2) Heal & Son. 94, swift, Whitney Museum of American Art, New York. 94, drying rack, Woldbye & Klemp. 95, music stand (2), Woldbye & Klemp. 96, Sciangai (2), Masera. 97, rulers, Woldbye & Klemp. 100, Optic Wonder, Woldbye & Klemp. 101, army compass, Woldbye & Klemp. 103, MacArthur, Popperfoto, Northampton. 104, tripod seat (2), Woldbye & Klemp. 105, Fiskars spade, Woldbye & Klemp. 105, u-dig-it, Woldbye & Klemp. 106, cabin lamp, Schnakenburg & Brahl. 107, Workmate (2), Woldbye & Klemp. 109, brushes (2), Woldbye & Klemp. 111, skate-scooter (2), Woldbye & Klemp. 115, Gannet, Jacob Fløche. 118, Opinel, Woldbye & Klemp. 120, Structura, Woldbye & Klemp. 123, Swiss Army knife, Woldbye & Klemp. 124, Leatherman (2), Woldbye & Klemp. 125, Gerber Multi-Plier (3), Woldbye & Klemp. 126, newspaper stick, Woldbye & Klemp. 128, electric cable, Woldbye & Klemp. 128, Gaston (2), Per Mollerup. 129, didactic roll-up, Woldbye & Klemp. 129, roller blind, Nationalmuseet. 130, Churchill, Archive Photo, Paris. 130, Borsalino, Woldbye & Klemp. 131, Smart Fun (2), Woldbye & Klemp. 131, Flexi-cal (2), Woldbye & Klemp. 132, weight measuring tape, Woldbye & Klemp. 132. measuring tapes, Woldbye & Klemp. 133, wire lock, Woldbye & Klemp. 133, Viper, Piotr & Co. 134, Victor Borge, Scanpix. 135, telescope, Woldbye & Klemp. 136, photo lamps, Woldbye & Klemp. 136, car aerial, Søren Kvistgaard. 137, crane lift, Jørgen Schytte. 138, Leica, Woldbye & Klemp. 139, Olympus (2), Woldbye & Klemp. 139, Minox (3), Woldbye & Klemp. 142, silver mug, Woldbye & Klemp. 142, plastic mug, Woldbye & Klemp. 142, slide rule, Woldbye & Klemp. 144–45, matryoshka dolls, Woldbye & Klemp. 147, brushes, Fabian Björnstjerna. 148, Klubi glasses, Marrku Alatalo. 148, dessert moulds, Woldbye & Klemp. 149, tea set (2), Woldbye & Klemp. 150, shopping baskets, Jørgen Schytte. 151, shopping trolleys, Jørgen Schytte. 151, chute, Jørgen Schytte. 152, Tivoli balloons, Jørgen Schytte. 154, hot-air balloons, Scanpix, 155, advertising balloon, John McIntyre. 156, inflated globes, Woldbye & Klemp. 157, traveller's cushion, Woldbye & Klemp. 157, air mail, Woldbye & Klemp. 157, washing-up bowl, Woldbye & Klemp. 157, coat hanger, Woldbye & Klemp. 157, beer cooler, Søren Kvistgaard. 159, self-inflating bed (2), Woldbye & Klemp. 159, sailor's mate, Scanpix. 160, tropical island, Scanpix. 161, playground, Jørgen Schytte, Billedhuset. 164, fan hat (3), Woldbye & Klemp. 164, colour swatches (2), Woldbye & Klemp. 164, allen keys, Woldbye & Klemp. 165, pantomime theatre, Per Mollerup. 166, television studio, BBC Picture Archive. 168, laundry rack, Per Mollerup. 169, paper sorter, Woldbye & Klemp. 170, disk cabinet (2), Woldbye & Klemp. 176, Wanscher stool, Aage Lund. 176, Klint stool, Kofoed. 176, Kjærholm stool, K. Helmer-Petersen. 176, Gammelgaard stool, Schnakenburg. 177 Koch stool, Torben Christian. 177, Aalto stool, The Museum of Modern Art. 178, Hitchcock, Polfoto. 180, chair bought by Koch, Schnakenburg. 180, Koch chair rack, Mogens S. Koch. 181, Aulenti chair, Masera. 183, outdoor chairs, Søren Kvistgaard. 183, Celestina (2), Masera. 184, Wegner chair, Erik Hansen. 185, anonymous chair, Schnakenburg. 186, beach chairs, Viggo Rivad, Billedhuset. 188, Klint deck chair, Schnakenburg. 189, Safari chair, Kofoed. 190, Blow, Ramazotti & Stucchi. 193, Magnussen (2), Sten Rønne. 194, Bush chair, Woldbye & Klemp. 194, Butterfly chair, Woldbye & Klemp. 195, bike with chairs, AFP, Polfoto. 196, restaurant chairs, Jørgen Schytte. 197, 3107, Strüwing. 199, Louis XX (2), Hans Hansen. 200, Spartana, Masera. 201, Santachair (2), Hans Hansen. 201, Miss C.O.C.O. (2), Matthew Donaldson. 203, Louisiana, K. Helmer-Petersen. 204, Kjærholm, K. Helmer-Petersen. 204, Ishøj Theatre (2), Schnakenburg & Brahl. 208, Boligens Byggeskabe, Mogens S. Koch. 209, Camilla, Masera. 210, Shaker table (2), Whitney Museum of American Art, New York. 214, Move It, Alexander Tröler. 215, Koch table, Mogens S. Koch. 216, La Loggia (2), Aldo Ballo. 220, Poker, Aldo Ballo. 221, PK54, K. Helmer-Petersen. 221, Bernt table, Brahl. 222, collapsible table, Lind, Ingemann. 222, Cumano, Aldo Ballo. 223, Aalto nest, Pietinen. 223, Rams nest, ingeborg Kracht-Rams. 223, Kjærholm nest, K. Helmer-Petersen. 223, Bernt nest, Schnakenburg. 226, Joy shelves, Ramazzotti. While every attempt has been made to trace copyright holders, the publishers would be grateful to be notified of any omissions, so that these may be included in any reprint.